RESEARCH IN MARITIME HISTORY
NO. 52

WAR AND TRADE IN EIGHTEENTH-CENTURY NEWFOUNDLAND

Olaf U. Janzen

International Maritime Economic History Association

St. John's, Newfoundland
2013

ISSN 1188-3928
ISBN 978-1-927869-02-4

Back issues of *Research in Maritime History* are available:

No. 1 (1991) David M. Williams and Andrew P. White (comps.), *A Select Bibliography of British and Irish University Theses about Maritime History, 1792-1990*

No. 2 (1992) Lewis R. Fischer (ed.), *From Wheel House to Counting House: Essays in Maritime Business History in Honour of Professor Peter Neville Davies*

No. 3 (1992) Lewis R. Fischer and Walter Minchinton (eds.), *People of the Northern Seas*

No. 4 (1993) Simon Ville (ed.), *Shipbuilding in the United Kingdom in the Nineteenth Century: A Regional Approach*

No. 5 (1993) Peter N. Davies (ed.), *The Diary of John Holt*

No. 6 (1994) Simon P. Ville and David M. Williams (eds.), *Management, Finance and Industrial Relations in Maritime Industries: Essays in International Maritime and Business History*

No. 7 (1994) Lewis R. Fischer (ed.), *The Market for Seamen in the Age of Sail*

No. 8 (1995) Gordon Read and Michael Stammers (comps.), *Guide to the Records of Merseyside Maritime Museum, Volume 1*

No. 9 (1995) Frank Broeze (ed.), *Maritime History at the Crossroads: A Critical Review of Recent Historiography*

No. 10 (1996) Nancy Redmayne Ross (ed.), *The Diary of a Maritimer, 1816-1901: The Life and Times of Joseph Salter*

No. 11 (1997) Faye Margaret Kert, *Prize and Prejudice: Privateering and Naval Prize in Atlantic Canada in the War of 1812*

No. 12 (1997)   Malcolm Tull, *A Community Enterprise: The History of the Port of Fremantle, 1897 to 1997*

No. 13 (1997)   Paul C. van Royen, Jaap R. Bruijn and Jan Lucassen (eds.), *"Those Emblems of Hell"? European Sailors and the Maritime Labour Market, 1570-1870*

No. 14 (1998)   David J. Starkey and Gelina Harlaftis (eds.), *Global Markets: The Internationalization of The Sea Transport Industries Since 1850*

No. 15 (1998)   Olaf Uwe Janzen (ed.), *Merchant Organization and Maritime Trade in the North Atlantic, 1660-1815*

No. 16 (1999)   Lewis R. Fischer and Adrian Jarvis (eds.), *Harbours and Havens: Essays in Port History in Honour of Gordon Jackson*

No. 17 (1999)   Dawn Littler, *Guide to the Records of Merseyside Maritime Museum, Volume 2*

No. 18 (2000)   Lars U. Scholl (comp.), *Merchants and Mariners: Selected Maritime Writings of David M. Williams*

No. 19 (2000)   Peter N. Davies, *The Trade Makers: Elder Dempster in West Africa, 1852-1972, 1973-1989*

No. 20 (2001)   Anthony B. Dickinson and Chesley W. Sanger, *Norwegian Whaling in Newfoundland: The Aquaforte Station and the Ellefsen Family, 1902-1908*

No. 21 (2001)   Poul Holm, Tim D. Smith and David J. Starkey (eds.), *The Exploited Seas: New Directions for Marine Environmental History*

No. 22 (2002)   Gordon Boyce and Richard Gorski (eds.), *Resources and Infrastructures in the Maritime Economy, 1500-2000*

No. 23 (2002)   Frank Broeze, *The Globalisation of the Oceans: Containerisation from the 1950s to the Present*

No. 24 (2003)   Robin Craig, *British Tramp Shipping, 1750-1914*

No. 25 (2003)   James Reveley, *Registering Interest: Waterfront Labour Relations in New Zealand, 1953 to 2000*

Research in Maritime History would like to thank Memorial University of Newfoundland for its generous financial assistance in support of this volume.

Table of Contents

i

ABOUT THE AUTHOR

OLAF U. JANZEN <olaf@grenfell.mun.ca> is Professor of History at the Grenfell Campus of Memorial University of Newfoundland in Corner Brook, Newfoundland. He is a Fellow of the Royal Historical Society and a member of a number of organizations, including the International Maritime Economic History Association, the Canadian Nautical Research Society, the Society for Nautical Research, the Navy Records Society and the Newfoundland Historical Society. Dr. Janzen's research has focussed on elements of the defence, society and trade of eighteenth-century Newfoundland, and he has published frequently in peer-reviewed journals, including the *International Journal of Maritime History* and *Newfoundland and Labrador Studies*. He is the author of an on-line "Reader's Guide to the History of Newfoundland and Labrador to 1869" <http://www2.grenfell.mun.ca/nfld_history/index.htm>.

Introduction

I am flattered to have been invited to reprint a selection of my papers over the course of nearly thirty years, and I am grateful to the series editor of *Research in Maritime History* and to the International Maritime Economic History Association for encouraging me to accept that invitation.

The papers presented in this volume are not arranged in the order in which they were written and published. Rather, I have placed them in an approximate order of historical chronology in the belief that this will give the volume greater coherence and value. Allow me therefore to explain how they came to be written, and perhaps why.

My interest in eighteenth-century Newfoundland began as a graduate student at Queen's University in Kingston, Ontario, Canada. I was drawn there in 1971 by George Rawlyk who had published on the concept of "revolution rejected." This refers to the fact that, while many of the British American colonies declared their independence from Great Britain in 1776, a few – either through circumstance or choice – did not. The result was a graduate essay on Newfoundland's response to the American Revolution. At the same time, I came under the tutelage of Donald Schurman, who was then also teaching at Queen's in the field of military and naval history and theory. These were serendipitous developments, for Rawlyk more than any other was responsible for my application to Memorial University's Corner Brook campus, where I have worked to this day, while Schurman more than any other was responsible for mentoring what skills I can claim as an historian. Both served as supervisors of my doctoral dissertation.

That dissertation expanded on the graduate paper I had written for George Rawlyk. My original intention had been to evaluate the nature and effect of American privateering activities in Newfoundland waters during the American War of Independence. I conceived it as a Newfoundland equivalent to John Dewar Faibisy's dissertation on American privateers in Nova Scotia during that same conflict.[1] It had been Faibisy's contention that American privateers disrupted communications and movement sufficiently in Nova Scotia to impair that colony's ability to develop a capacity, one way or another, to respond in a united way to the issues of the American Revolution. Might this also have been the case in Newfoundland?

---

[1]John Dewar Faibisy, "Privateering and Piracy: The Effects of the New England Raiding Upon Nova Scotia During the American Revolution, 1775-1783" (Unpublished PhD thesis, University of Massachusetts, 1972).

1

This focus guided my first overseas research foray to the British archives, and it was still my focus when I subsequently travelled to Newfoundland to pursue my research further at the provincial archives in St. John's. But then I met Keith Matthews, who had recently completed his own dissertation on Newfoundland's economic and social connection with England's West Country.[2] Matthews seemed supportive of my own focus at first, but quickly suggested that there wasn't enough source material available to support a doctoral dissertation, and that a broader question that should be – and could be – answered involved Newfoundland's place in British strategic responses to the American insurrection. I had already come privately to believe that my original topic was not working out, and I realized, based on the work I had already done, that Matthews was right – a strategic analysis would work. The result was "Newfoundland and British Maritime Strategy during the American Revolution."[3]

That dissertation became the source for at least three of the papers that appear in this volume. Ironically, the first of these was the last to appear – "The Royal Navy and the Defence of Newfoundland During the American Revolution." It had been presented in 1984 at my first-ever historical conference and subsequently published in what is arguably one of the finest historical journals in Canada, *Acadiensis*, a publication devoted to the history of Atlantic Canada under the guiding hand at the time of one of its best editors, David Frank.[4] The next year, I attended the Seventh Naval History Symposium at the US Naval Academy in Annapolis, Maryland, where I presented a paper on "The French Raid upon the Newfoundland Fishery in 1762." Subtitled "A Study in the Nature and Limits of Eighteenth Century Sea Power," this paper benefited immeasurably from the conference commentary by Daniel A. Baugh and was selected as one of the symposium papers to be published in 1988 as a representative sampling of that conference. A third paper reprinted here, "The American Threat to the Newfoundland Fisheries, 1776-1777," was also published in 1988 but had its gestation first as chapters in my dissertation and then

---

[2]Keith Matthews, "A History of the West of England-Newfoundland Fisheries" (Unpublished DPhil thesis, Oxford University, 1968). This dissertation has since been digitized by the Centre for Newfoundland Studies at Memorial University and is available on-line through the MUN Library system as a pdf file: <http://collections.mun.ca/PDFs/cns/MatthewsKeith.pdf>.

[3](Unpublished PhD thesis, Queen's University, 1984).

[4]The conference where the paper was presented was the "Bermuda/Canada 1609-1984 Conference" held in Bermuda in 1984. It was an exciting and rewarding debut for a young historian, but my colleagues still raise eyebrows at the notion of an academic seeking funding to travel to a conference in Bermuda at a time when the campus where he is employed was in the subarctic grip of a Newfoundland winter.

as a paper presented in 1985 at a colloquium sponsored by the Maritime History Group at the St. John's campus of Memorial University of Newfoundland with the then-title "A Rage for Profit? The Abortive American Raids on the Newfoundland Fisheries, 1776 and 1777."

By then, my employment at Memorial University's campus on Newfoundland's west coast had begun to draw me into new research interests, though still within the broad framework of eighteenth-century Newfoundland. The region had received very little scholarly attention, and this alone piqued my interest. As well, my work on the French raid of 1762 had given me an appreciation for the degree to which Newfoundland history had been shaped not just by the British but also by the French. Specifically, my research by then had driven forcefully home the realization that far too much of the existing literature on eighteenth-century Newfoundland was written from a predominately (if not exclusively) Anglophone perspective. The French documentation relevant to that history was rich and abundant, and much of it was being used by historians such as Jean-François Brière and Laurier Turgeon in North America and a host of others in France. I soon found myself drawn into a study of French activities in southwestern Newfoundland which included fishing, trade and even settlement. A visit to the research archives at Fortress Louisbourg National Historic Site confirmed that there had been a great deal of interaction between the French colony of Île Royale and the nearby coasts of southwestern Newfoundland even though, by the terms of the Treaty of Utrecht, the French were not supposed to fish in that part of Newfoundland, let alone settle there. This led to some preliminary assessments and conclusions about those activities, presented at a 1987 conference and subsequently published in the journal *Newfoundland Studies*.[5] The results were sufficiently promising to make this the focus of more intensive research in the UK and France during a sabbatical and eventually to the second paper on that topic, "'Une petite Republique' in Southwestern Newfoundland," which is reprinted here.

Like most of my publications, this research reflected an interest in both the particular – in this case the evolution of a specific community in eighteenth-century Newfoundland and the European merchants whose commercial links with other parts of the North Atlantic enabled the settlement to develop – and the general – the degree to which communities in early modern Newfoundland existed not despite merchant objections (a common misperception in the literature) but because of merchant support. This last was not an original idea. As I admitted earlier, my work tends to build on the ideas and conclusions of others. In this case, historical geographers like John Mannion

---

[5]"'Une grande liaison:' French Fishermen from Île Royale on the Coast of Southwestern Newfoundland, 1714-1766 – A Preliminary Survey," *Newfoundland Studies*, III, No. 2 (1987), 183-200.

and Gordon Handcock, and anthropologists like Tom Nemec, had built on the foundation of Keith Matthews' historical revisionism to show how settlement in eighteenth-century Newfoundland would not have been possible without the transatlantic commercial linkages, credit and commercial services of merchants. The French who settled at Codroy in the 1730s and 1740s were not a departure from Newfoundland's settlement history but another expression of it.

I also argued in my Codroy papers that the community demonstrated the degree to which the jurisdictional authority of mercantilistic empires could be constrained by eighteenth-century realities. It is therefore something of an irony that the jurisdictional ambiguities and administrative indifference of Versailles and London towards people settling on the fringes of empire in southwestern Newfoundland worked only in peace time; when war broke out, the conditions which buffered the settlements from official interference would not protect them from the forces of war, as I explain in "Un Petit Dérangement: The Eviction of French Fishermen from Newfoundland in 1755." Though early versions were presented at conferences, this paper has never previously been published.

I continued to pursue the theme of imperial control – or the challenges of asserting it – in other papers. "Showing the Flag: Hugh Palliser in Western Newfoundland, 1764" appeared soon after "Une petite Republique" had been published. The unprecedented visit to the west coast in 1764 by Hugh Palliser, the naval commandant in Newfoundland that year who also exercised the function of the island's civil governor, had been touched on before by William Whiteley, but only as part of a broader study of Palliser's services in Newfoundland between 1764 and 1768.[6] I focussed specifically on that 1764 visit, arguing that it was a belated response by British authorities to the way in which a French presence at Codroy during the decades before the Seven Years' War had weakened, even challenged, British pretensions of sovereignty over that region. Yet assertions of sovereignty could also be made through cartographic initiatives, as I argued both in that paper and in another which appeared a few years later, "'Of Consequence to the Service:' The Rationale Behind Cartographic Surveys in Early Eighteenth-Century Newfoundland." Both papers reflect my continuing interest in the ways in which empires increasingly acted to consolidate and articulate their authority over regions where a more casual approach had previously been acceptable. For the British, the services that could be rendered by the ships of the Royal Navy stationed in Newfoundland waters was crucial to this process.

My Codroy research invited exploration of other themes as well. One concerned aboriginal migration from Nova Scotia to southwestern Newfoundland in the 1760s. This led to a co-authored paper which investigated why,

---

[6]William Whiteley, "Governor Hugh Palliser and the Newfoundland and Labrador Fishery, 1764-1768," *Canadian Historical Review*, L, No. 2 (1969), 141-163.

how and when that migration occurred.[7] Years later, I returned to that theme in a paper which focussed more on the efforts of the Royal Navy in peacetime to discourage that migration – "The Royal Navy and the Interdiction of Aboriginal Migration to Newfoundland, 1763-1766," a paper which is also reprinted in this volume. A second theme emerged from a further exploration into the illicit comings and goings of the Chenu brothers, part of a merchant family of Saint-Malo, whom I had linked with the creation and growth of the Codroy settlement. When the opportunity materialized to engage in the lucrative but illicit movement of British cod into Spain at a time when the Spanish market was closed to direct imports of British fish, the Chenus responded vigorously, as I explain in another paper reprinted here, "The Illicit Trade in English Cod into Spain, 1739-1748." A third theme was that of piracy in Newfoundland waters. This too was a theme which had first been raised by the Codroy settlers because when British authorities first complained that France was supporting settlement in southwestern Newfoundland, French authorities had responded by insisting that the settlements were beyond their jurisdiction and control, and that the people living there were brigands and pirates, as much a nuisance to France as to England. Were they pirates? No, of course not, as I made clear in my papers on those settlements. But the accusation of piracy was plausible because following the conclusion of the War of the Spanish Succession in 1713, there had been a resurgence of piracy in the North Atlantic, and my interest in this sharp but short-lived problem resulted in yet another paper reprinted here on "The Problem of Piracy in the Newfoundland Fishery in the Aftermath of the War of the Spanish Succession." Like so many of my papers, this one developed an interpretation which placed particular details (in this case, about the visit of pirate Bartholomew Roberts to Trepassey) within a broader context of North Atlantic piracy which had been developed by Markus Rediker, Robert Ritchie and others.

Of all the papers reprinted here, the one which appears least related to the others was my study of "A Scottish Sack Ship in the Newfoundland Trade, 1726-27," based on research conducted in Edinburgh during a half-sabbatical in early 1995 and subsequently published in 1998 in *Scottish Economic and Social History*. The Scots were not major players in the Newfoundland fishery and fish trade for most of the eighteenth century. Yet in developing the several themes already described here, I stumbled across references to a Scottish ship which carried Newfoundland fish to Spain in 1727. Again my interest was piqued, sufficiently so that I resolved to learn more. The result was an abundance of documentation, some of which supported the paper published here and all of which I still hope can be shaped eventually into a more detailed publication.

---

[7]With Dennis A. Bartels, "Micmac Migration to Western Newfoundland," *Canadian Journal of Native Studies*, X, No. 1 (1990), 71-94.

All of this work, together with my academic teaching, challenged me to become as familiar as possible with as much of the literature on early modern Newfoundland as possible. One serendipitous result of this obsession with the literature was an invitation to contribute a chapter on "Newfoundland and the International Fishery" to an annotated bibliographical guide to Canadian history.[8] I subsequently added and greatly expanded that essay on-line.[9] That work, in turn, introduced me to maritime history and its full diversity. While it is a world in which I claim no specific expertise outside of eighteenth-century Newfoundland, I do believe it has made me well informed within a broad range of specializations – economic, social, naval, diplomatic – to the point that I was invited to present a paper at the World Marine Millennial Conference held in Salem, Massachusetts in 2000. That paper, "'A World Embracing Sea:' The Oceans as Highway, 1604-1815," is also reprinted in this volume.

All in all, it has been a thoroughly pleasurable journey, from graduate studies to academic career, and from my first hesitant steps into Newfoundland history into a much fuller appreciation of the degree to which Newfoundland's history cannot be properly understood in isolation from other parts of the North Atlantic world. Best of all, the journey has introduced me to a host of academics on both sides of the Atlantic, many of whom have become friends, and all of whom have been extremely helpful to me in my efforts to understand better the tremendous complexity that is Newfoundland history in the eighteenth century. I am most grateful to them all.

---

[8]"Newfoundland and the International Fishery," in M. Brook Taylor (ed.), *Canadian History: A Reader's Guide. Vol. I: Beginnings to Confederation* (Toronto, 1994), 280-324.

[9]"A Reader's Guide to Newfoundland History to 1869" <http://www2. grenfell.mun.ca/nfld_history/index.htm>.

"A World Embracing Sea:"
The Oceans as Highway, 1604-1815[1]

From the passing of the Elizabethan Age at the beginning of the seventeenth century to the end of the Napoleonic wars just after the end of the eighteenth century, European commerce began to flow along oceanic highways that were truly global in extent. It was a period of history that encompassed an enormous number of experiences, developments and nationalities. Just in European terms alone, the "Oceans as Highways" embraced the Dutch, French, Spanish, Portuguese and English maritime experiences, as well as those of Sweden, Denmark, Russia, the United States and a host of smaller nationalities, city-states and principalities. It was an age characterized by what historians today refer to as "mercantilism" in which maritime trade was dominated by a rapidly expanding volume of luxury and consumer commodities such as sugar, coffee, tobacco, fish and grain. It was equally an age when oceanic highways permitted the rapid growth of the traffic in African slaves to the New World and when European powers struggled increasingly – not only on land but also on the sea itself – to control and use the wealth that travelled on these oceanic highways. To say something reasonably coherent about such a diversity of material within a brief chapter is a daunting challenge. Therefore, I will explore several themes within the narrower framework of the Newfoundland fishery, fisher society and fish trade during the eighteenth century. Like so many other oceanic activities, this is one in which all aspects of oceanic commerce are manifest: oceans as highways; the struggle for control of the commerce on those highways; the national and ethnic diversity of the shipping, markets and individuals engaged in oceanic commerce; and above all the strategies and considerations necessary to succeed in oceanic commerce. Through this admittedly narrow window into the past, it is possible to explore the way in which the oceans functioned as highways for ships, people and commodities during the seventeenth and eighteenth centuries.

Before the seventeenth century, Europeans had already developed significant maritime highways as they engaged in commerce within and beyond Europe. The Baltic, the Mediterranean, the English Channel and the Atlantic seaboard had all become busy with shipping and trade. During the sixteenth century, Europeans exploded beyond the confines of their homelands as they extended their reach to Africa, Asia and the New World. By the seventeenth

[1]This essay appeared originally in Daniel Finamore (ed.), *Maritime History as World History: Proceedings of the World Marine Millennial Conference 2000* (Gainesville, FL, 2004), 102-114.

and eighteenth centuries, trade was expanding along well-established routes across the Atlantic to North and South America, south to Africa, across the Indian Ocean to India, Southeast and East Asia, and then back again on homeward voyages. A broad range of commodities flowed along these oceanic highways – sugar from the West Indies; tobacco from the Chesapeake; fish from Newfoundland; tea, silks and porcelain from China; and spices from the East Indies. A great diversity of manufactured goods went to overseas destinations, as did thousands of migrants in search of new opportunities and new homes, and slaves were carried by the millions out of Africa to both the New and Old Worlds. The ships engaged in these trades were as diverse in size and appearance as the cargoes they carried, from large, well-manned ships of the British East India Company to the small fifty- or seventy-ton vessels employed in the fisheries. Nevertheless, for the most part they evolved very little after 1700, the major refinements having been worked out in the course of the seventeenth century – a mariner of 1700 would have had no difficulty working a ship of 1800. He might have observed that the prominent quarterdecks and fo'c'sles characteristic of ship profiles in the seventeenth century had gradually diminished so that a great many vessels were flush-decked by the late eighteenth century. He would have noticed that the seventeenth-century tiller had given way to the wheel, and he might have realized that copper sheathing had begun to appear on the bottoms of ships that ventured into the tropics, particularly on those ships that specialized in the slave trade.[2]

Until well into the seventeenth century, most ships engaged in oceanic shipping were Dutch. By one calculation, the Dutch owned more tonnage than the rest of Atlantic Europe combined. This domination is typically explained with reference to the development of the *fluit* late in the sixteenth century which gave the Dutch the ideal oceanic cargo-carrier – ships that, according to one observer, "measure little and stow much."[3] It is, on the face of it, a reasonable argument; the *fluit* may not have been designed for speed, but it was ideally suited for the movement of bulk cargoes, where transportation costs were generally greater than the costs of production. Its small crew enabled the Dutch to earn more profit per ton of vessel than anyone else. Yet surely all this begs the question of what enabled the Dutch provinces to build so many ships.

---

[2]David Williams, "The Shipping of the British Slave Trade in Its Final Years, 1798-1807," *International Journal of Maritime History*, XII, No. 2 (2000), 20, esp. table 7. Using a sampling of ships employed in the slave trade at the turn of the nineteenth century, Williams determined that more than ninety-three percent were sheathed, most with copper. In contrast, in other trades only ten percent was copper-sheathed in 1806.

[3]Cited in Ralph Davis, *The Rise of the English Shipping Industry in the Seventeenth and Eighteenth Centuries* (London, 1962; reprint, St. John's, 2012), 46.

Clearly, innovations in shipping technology are not, in and of themselves, sufficient explanation to account for predominance on the oceanic highways.

Eventually, other nations acquired Dutch *fluits* or developed designs of their own that were patterned in part on the *fluit*.[4] By the middle of the seventeenth century, they were also introducing policies designed to liberate themselves from the Dutch stranglehold on shipping and commerce, policies to which historians later attached the word "mercantilism." By the eighteenth century, both France and England relied far more on their own shipbuilding capabilities than on those of the Dutch who, in a relative sense at least, began to fade as the dominant maritime commercial power. Yet this, too, suggests that the key to understanding changing patterns of trade and domination at sea rests with our ability to recognize the significance of what was happening ashore.

Because this was an age of sail, shipping routes and even schedules were determined in considerable measure by factors over which ships had little control, but to which they had instead to adapt, such as oceanic currents or prevailing patterns of winds. Europeans had learned by experience that in order to cross the Atlantic as expeditiously as possible, the best route did not necessarily follow the shortest line on the chart. Indeed, the winds that blew from Europe to the West Indies and the northern coast of South America did so with such predictability that maritime trade could depend on these winds for a reliable and reasonably predictable passage – hence the name "trade winds." The course followed by a merchantman making its way from the West Indies to Europe took it north to the latitude of Bermuda, then northeast until it reached the great fishing banks south of Newfoundland where the ship would catch the westerlies for the Atlantic crossing. For a merchantman making its way from Europe to North America, however, the westerlies were a costly inconvenience that required tacking and wearing on a zig-zag course for several weeks. Consider, for instance, the seventy-ton *Christian* which departed Leith in Scotland in June 1726 on a voyage to Newfoundland with a load of biscuit; the passage across the Atlantic to St. John's took six weeks, roughly twice the time taken by English West Country ships that made the crossing earlier in the season before the prevailing westerlies of summer could hinder their crossing.[5] In 1746, also in the month of June, a naval expedition that made its way from France to Nova Scotia followed a course that was longer in distance than *Christian*'s route. By sailing south from its home ports before

---

[4]Roger Morris, *Atlantic Seafaring: Ten Centuries of Exploration and Trade in the North Atlantic* (Camden, ME, 1992), 112-113. The lineage of eighteenth-century pinks and cats can be traced back to the *fluit*, according to Morris.

[5]Olaf U. Janzen, "A Scottish Venture in the Newfoundland Fish Trade, 1726-1727," in Janzen (ed.), *Merchant Organization and Maritime Trade in the North Atlantic, 1660-1815* (St. John's, 1998), 140.

venturing across the Atlantic, however, the expedition attempted to catch the trade winds so that the sail-handling skills of its inexperienced crews would not be overly challenged.[6]

Winds were not, however, the only consideration in selecting a course on the oceanic highways of the day. There were also ocean currents such as the Gulf Stream. While their precise nature was not then fully understood, their effect on the movement of ships on the oceanic highways was known to all, and ships either avoided a course that would oblige them to sail against the current or chose one that enabled them to pick up additional speed by sailing with the current. The capabilities of the ship itself also mattered – its ability to sail well into the wind, to carry cargo and to be worked by an economically sized crew. The anticipated length of the voyage was also a consideration.

By reason of the roundabout routes that the oceanic highways followed, as well as for very sensible reasons relating to the constant search to maximize the profits of a voyage, ships and vessels engaged in the Atlantic trades developed distinct three-point voyages that gave rise to the idea of "triangle trades." The term is a simplification of the way in which ships found it more profitable to add legs to their voyage pattern. For instance, a ship might carry a cargo of wines and fabrics from France to the West Indies, then transport rum and molasses – and perhaps a reserved portion of the wine and fabrics – to Louisbourg or Canada before returning to France with furs, fish or timber. Conversely, a ship might carry manufactured goods and wines from France to Canada, and then pick up a cargo of fish and wood and deliver it to the West Indies before returning to France with sugar. Ideally, each leg of the voyage should make a profit. With three legs per year, a ship would generate more profit than a vessel engaged in a bilateral trading pattern. It was this kind of logic that inspired the Scottish merchants who chartered *Christian* to venture into the Newfoundland trade in 1726. They already had some experience exporting Scottish fish to pay for the Iberian wines they imported. The idea of selling Scottish biscuit in Newfoundland in exchange for fish that could then be carried to Spain and exchanged for cork and wine seemed a reasonable and profitable extension of their trade.[7]

The notion that many ships followed a three-point pattern over the course of a year can be something of an oversimplification. For instance, in the Newfoundland trade the vessels that carried men from England to the fish-

---

[6]James Pritchard, *Anatomy of a Naval Disaster: The 1746 French Expedition to North America* (Montréal, 1995), chap. 4. The expedition sailed south to catch the Portuguese trade winds that carried them to the northeast trades as far as the mid-Atlantic, where a change in course could be made to Nova Scotia. Unfortunately, contrary and variable winds dragged the voyage out; the expedition was at sea for three months instead of the expected seven weeks before it could anchor in Nova Scotia.

[7]Janzen, "Scottish Venture," 136-139.

ery were not necessarily the same vessels that carried the fish from Newfoundland to southern Europe. By the late sixteenth century, vessels that played no
role in the process of catching or curing fish, but instead did nothing more
than move cargoes of fish to market, had made their appearance.[8] Moreover, a
merchant-venturer engaged in the Newfoundland trade did not necessarily con-
fine his activity to fish. Rather, he might invest in several commodities so that
his ship might carry several cargoes to various destinations before it made its
way back to its point of origin in Europe. For instance, a ship arriving in
Spain with a cargo of Newfoundland fish might take on a mixed cargo of Iberian fruit, wines, ironwork and silks to be sent to America and exchanged for a
cargo that might take the ship back to Newfoundland, or that might take it on
to the West Indies before heading home. The term "triangle trade" should not,
therefore, suggest that the precise movement of a given ship in a particular
commodity trade can be predicted. Nevertheless, there is some validity to the
term, for it accurately describes the principal directions in which investment
capital in the form of shipping and cargoes moved.[9]

One advantage of such a shipping pattern is that it put a ship to profitable use practically the entire year round. The drawback was that a triangular
voyage pattern also took more time than a bilateral one, with the result that
oceanic trade was constantly dominated by deadlines that must have contributed significantly to the anxiety of the merchant-venturer. Failure to meet
those deadlines could defeat the voyage, perhaps even discourage investors
from venturing further onto the oceanic highways. This, too, can be demonstrated with reference to the voyage of *Christian*, although it must be understood that the same concerns existed in most trades. *Christian* was fitted out in
May and departed in June. It should have made Newfoundland in July but arrived instead in August, by which time the purchase price of fish had been set
at a higher level than would have been the case with an earlier arrival. The
ship acquired its cargo of fish quickly enough to make its departure for Spain
in September, but it was a slow sailer, so that by the time it reached Barcelona
the demand for its fish had already been satisfied by those who arrived before.
This made it difficult to dispose of *Christian*'s cargo and therefore imposed
additional delays on the ship. By the time a return cargo of wine had been secured, the season was so advanced that the prevailing Mediterranean winds

---

[8]Peter Pope, "Sack Ships in the Seventeenth-Century Newfoundland Trade,"
Northern Seas Yearbook 1999 (St. John's, 2001), 33-46.

[9]Pauline Croft, "English Mariners Trading to Spain and Portugal, 1558-
1625," *Mariner's Mirror*, LXIX, No. 3 (1983), 257. Croft identifies *Parnell* of London
(1579-1580) as "one of the earliest ships to have completed the new triangular voyage
from England to...Newfoundland...and thence down to the Mediterranean." Gillian
Cell describes the triangular trade in *English Enterprise in Newfoundland, 1577-1660*
(Toronto, 1969), 31-32.

now blew contrary, denying the crew an easy return voyage to Gibraltar. Just as *Christian* arrived there, long-standing tensions between Spain and England caused hostilities to break out; *Christian* and all other shipping at Gibraltar were forced to wait nearly a month before a convoy could be arranged for their safety. More than a year passed between *Christian*'s departure from Leith and its return home. Whether the voyage earned a profit at all is moot; there is certainly no evidence that the merchants who chartered *Christian* ever tried to invest in the Newfoundland trade again.[10] There may have been a pronounced rhythm to many of the commodity trades that employed the oceanic highways, but none could be described as a simple business.[11]

Given their duration, oceanic voyages could present a ship with many risks and hazards, so that those who ventured forth onto the oceanic highways were understandably preoccupied with finding ways to reduce risks. In 1604, *Hopewell* of London took out insurance against the following hazards of the Newfoundland trade: "the seas men of warre, Fire, enemies, Pirates, Robers, Theeves, Jettesons, Letters of Mark and counter Mark, Arrestes, Restraintes, and detaynements...barratrye of the Master and Marriners of all other perilles, losses and misfortunes whatsoever they be."[12] The nature of most trades generally, and the gaps in the communication between the owner and the master of merchantman in particular, meant that the master had to be given considerable freedom to exercise independent judgment. Trust was, therefore, an essential ingredient in the working relationship between shipmaster and merchant in oceanic trades. Many merchants encouraged younger members of the family to serve as shipmasters, in part to assure themselves of a reliable servant, in part to apprentice a possible heir in the business. A particularly successful shipmaster might be invited into a partnership role, a relationship that might then be further cemented by a carefully arranged marriage into the family.

In France, for instance, an important business practice was *la société familiale*. This refers to the intricate network of shared investment and vessel ownership by which family connections were used to secure French businesses against the many risks of eighteenth-century commerce.[13] Under this arrange-

---

[10]Janzen, "Scottish Venture," 140-153.

[11]A very perceptive analysis of the deadlines that had to be met by a French ship engaged in the Newfoundland fishery and trade is provided in Jean-François Brière, "Le commerce triangulaire entre les ports terre-neuviers français, les pêcheries d'Amérique du nord et Marseille au 18e siècle: nouvelles perspectives," *Revue d'histoire de l'amérique française*, XL, No. 2 (1986), 193-214.

[12]Cell, *English Enterprise*, 12. "Barratrye" or "barratry" refers to fraud or gross negligence of a ship's officer or seamen against a ship's owners or insurers.

ment, all members of a family would contribute their personal capital into a common family fund for purposes of owning, outfitting and crewing several vessels and ships. Business capital and family fortunes blended and became indistinguishable. The family, in effect, became a joint shareholding venture, with the profits of the voyages distributed among the members of the family according to the amount each had contributed. In this way, the security of share ownership and business partnership was cemented by blood-ties and kinship. The participants in this kind of family-business might relocate in several seaports, and specialize in different activities, all to ensure the well-being of a family involvement in oceanic commerce.

An excellent example of the way family and kinship were used to secure investment in oceanic commerce is provided by the brothers Chenu of Saint-Malo during the first half of the eighteenth century: Claude Chenu, Sieur Boismory (ca. 1678-17?); Pierre Chenu, Sieur Dubourg (ca. 1683-1769); Jacques Chenu, Sieur Duchenôt (ca. 1687-1758); and Louis Chenu, Sieur Duclos (ca. 1694-1774). A fifth Chenu, Jerôme, Sieur Dupré (ca. 1698-17?) may have been a cousin.[14] The Chenus did not belong to the top rank of Saint-Malo's *négociants*; they lacked the wealth, diversity of commercial activity, and international associations that were definitive characteristics of the great merchants.[15] Rather, their business dealings were limited to the French North American cod fishery and its related activities – outfitting, shipowning and trade. As young men, they all served as captains of fishing or trading vessels that they either owned themselves or that were owned by a brother. This in

---

[13]The importance of family in eighteenth-century French business is discussed in André Lespagnol, "Une dynastie marchande malouine: les Picot de Clos-Rivière," Société d'histoire et d'archéologie de l'Arrondissement de Saint-Malo, *Annales 1985*, 233; and Laurier Turgeon, "Les échanges franco-canadiens: Bayonne, les ports basques, et Louisbourg, Île Royale (1713-1758)" (Unpublished mémoire de maîtrise, Université de Pau, 1977), 86.

[14]Information on the family has been compiled from various registers, reports, declarations and other manuscripts housed in the Archives of Fortress Louisbourg National Historic Park, Louisbourg, Nova Scotia; Archives départementales de l'Ille-et-Vilaine, Rennes; Archives de l'arrondissement maritime de Rochefort, Rochefort; and Archives départementales de la Charente-Maritime, La Rochelle. Also of great use was L'Abbé Paul Paris-Jallobert, *Anciens Registres Paroissiaux de Bretagne (Baptêmes - Mariages - Sépultures): Saint-Malo-de-Phily; Evêché de Saint-Malo - Baronnie de Lohéac - Sénéchaussée de Rennes* (Rennes, 1902); and Paris-Jallobert, *Anciens Registres Paroissiaux de Bretagne (Baptêmes - Mariages - Sépultures): Saint-Malo (Evêché - Seigneurie commune - Sénéchaussée de Dinan)*, vol. I (Rennes, 1898). The parents of the Chenu brothers were Jean Chenu and Margueritte Porée.

[15]André Lespagnol, *et al.*, *Histoire de Saint-Malo et du pays malouin* (Toulouse, 1984), 152-153.

itself was fairly typical of all levels of merchants engaged in overseas commerce, though Louis Chenu appears to have been content to remain a captain-owner throughout his later years, even as his brothers were establishing themselves as merchants of Saint-Malo or its suburb, Saint Servan. Claude Chenu was the most mobile of the four; at various times he was identified as a resident of Granville, Saint-Malo and La Rochelle, although he always maintained both the business and personal sides of his relationship with his brothers. Together the Chenus maintained fishing stations, owned vessels, sometimes as individuals but usually in partnership with one another, and participated in the truck trade at Île Royale (as Cape Breton Island was known to the French). Their sons provided them with a large labour pool from which were drawn the captains and junior officers of the Chenu vessels and the next generation of merchants. In short, they were a fairly typical example of a phenomenon visible within both the French and West Country English Newfoundland fisheries: a family trying to emerge out of the ranks of shipmasters and become merchants, shipowners and outfitters determined to emulate those in their community who, through their own success, had demonstrated that it was possible to rise from equally modest means to the highest ranks of Malouin commercial society.[16]

The reduction of risks and the safeguarding of investment in oceanic commerce were enhanced by factors and methods other than the bond of family – religion, for instance. Throughout the world of European oceanic trade, examples abound in which religious fellowship combined with family ties cemented through marriage to create powerful business alliances – for instance, the Huguenots of seventeenth-century La Rochelle or eighteenth-century Carolina, or the Quakers of eighteenth-century Newfoundland.[17] The element of trust, on which commercial credit depended so absolutely, was thus ensured in many informal ways.

Yet trust only went so far. Good old-fashioned influence was an essential ingredient in securing one's investments on the oceanic highways of the seventeenth and eighteenth centuries. Alan Pearsall has recently demonstrated

---

[16]*Ibid.*, 131-132. One such model of success was Noël Danycan, Sieur de l'Epine (1656-1734), the son of a small outfitter who began his own career as a captain-outfitter of a Newfoundland fishing vessel. Worth only 15,000 *livres* at the time of his marriage in 1685, he was worth millions twenty years later thanks to investments in the South Sea trade.

[17]John F. Bosher, "The Gaigneur Clan in the Seventeenth-Century Canada Trade," and R.C. Nash, "The Huguenot Diaspora and the Development of the Atlantic Economy: Huguenots and the Growth of the South Carolina Economy, 1680-1775," both in Janzen (ed.), *Merchant Organization and Maritime Trade*, 15-51 and 75-105; and W. Gordon Handcock, "The Poole Mercantile Community and the Growth of Trinity, 1700-1839," *Newfoundland Quarterly*, LXXX, No. 3 (1985), 19-30.

the way in which the Russia Company pressed the Admiralty for warships to escort and protect its trade with Archangel during the War of the Spanish Succession.[18] Merchants of the English West Country who engaged in the Newfoundland trade used their positions as mayors and aldermen of their communities to petition London for measures to protect their commercial interests. Of course, it can be difficult to determine precisely how effective such pressure actually was. Historians often assume that lobbying by the Newfoundland fishing interests was effective because the policies adopted by government seemed to match the policies desired by the trade. Yet government perceptions and priorities could differ significantly from those of the fishing interests, making it necessary for the merchants to word their petitions in such a way that their real objectives were disguised in order to assure themselves of a favourable response from government. Thus, the view that the Newfoundland fishery was a "nursery for seamen" has been challenged on the grounds that the fishery provided the navy with relatively few mariners and that its image as a "nursery" was an illusion maintained by the merchants to justify their resistance to government regulation of their industry.[19]

Nevertheless, there is little doubt that one of the most striking developments in oceanic trade during this period is the degree to which government legal, diplomatic and military measures shaped mercantile opportunities. This was made abundantly clear at a recent conference session on the theme "Merchant Organization and Maritime Trade in the North Atlantic, 1660-1815." Several historians explored a diversity of examples, including government-encouraged penetration of trade with Iceland by British merchants during the Anglo-Danish War of 1807-1814; the success of local merchants in wresting control over commerce in the Spanish port of Bilbao from foreign merchants in the late 1600s; and the sudden opportunity presented by the Act of Union of 1707 for Scottish merchants to break into British Atlantic trades that had been previously denied them by virtue of mercantilistic restrictions.[20] What is also

---

[18]Alan Pearsall, "The Royal Navy and the Archangel Trade, 1702-14," *Northern Seas 1996* (Esbjerg, 1996), 64.

[19]See Gerald Graham, "Fisheries and Sea Power," Canadian Historical Association, *Annual Report 1941* (Ottawa, 1941), 24-31, reprinted in G.A. Rawlyk (ed.), *Historical Essays on the Atlantic Provinces* (Ottawa, 1967), 7-16; and David J. Starkey, "The West Country-Newfoundland Fishery and the Manning of the Royal Navy," in Robert Higham (ed.), *Security and Defence in South-West England before 1800* (Exeter, 1987), 93-101.

[20]Anna Agnarsdóttir, " The Challenge of War on Maritime Trade in the North Atlantic: The Case of the British Trade to Iceland During the Napoleonic Wars," and Aingeru Zabala Uriarte, "The Consolidation of Bilbao as a Trade Centre in the Second Half of the Seventeenth Century," both in Janzen (ed.), *Merchant Organization and Maritime Trade*, 221-258 and 155-173; and Janzen, "Scottish Venture," 133-153.

striking throughout this period is the degree to which oceanic trade in the Atlantic, as opposed to the commerce with East Asia, the East Indies, Africa and even the West Indies did not depend on large-scale, corporate organization. There was no equivalent in the Atlantic trades of great merchant companies like the British, French and Dutch East India Companies, the Royal African Company or the Hudson's Bay Company. Was this because the state was better able to provide the measure of security that companies had to provide themselves in more remote or disputed waters? It is clear that more work is needed before the full complexity of the "oceanic highways" can be understood.

To conclude, there were many considerations that shaped behaviour on those highways – risk, trust, access to reliable information and an understanding of the nature and the limitations of government regulation, to name but a few. One thing is evident. Effective analysis of the use to which the oceanic highways were put during this period requires that we understand more than just the obvious needs of maritime commerce – a ship, a cargo and a crew. What takes place on land has as much a bearing on maritime history as what takes place at sea. To put it in other words, while it is undeniable that the oceans served as profoundly important highways during the seventeenth and eighteenth centuries, an understanding of the land-sea relationship is vital to our attempts to comprehend the nature of the oceanic highways.

"Of Consequence to the Service:"
The Rationale behind Cartographic Surveys in
Early Eighteenth-Century Newfoundland[1]

This essay examines two cartographic surveys carried out during the eighteenth century on the southern coast of Newfoundland. The surveys were conducted ostensibly for the benefit of trade and commerce. It is the contention of this paper, however, that of equal importance in the decision to commission the expeditions were questions of sovereignty. The two surveys – by William Taverner (1714-1715) and Lt. John Gaudy (1716) – were immediate responses to the dramatic changes that followed the Treaty of Utrecht (1713) by which France conceded British sovereignty over all of Newfoundland and gave up the right to maintain any permanent settlement there. As the French withdrew, the British moved swiftly to establish their authority in Placentia Bay, St. Pierre and the south coast of the island. The reluctance of British fishermen to move into the region, together with evidence that French fishing and commerce persisted there, contributed to the urgency of the surveys.

The issue of sovereignty in Newfoundland, and hence the desire for cartographic clarification, arose primarily because until the eighteenth century no single country claimed exclusive control over the island. Since its beginnings early in the sixteenth century, the Newfoundland fishery had been an international activity, attracting French, Portuguese, Basque, English and Dutch fishing and trading vessels.[2] By the second half of the seventeenth century, however, English and especially French fishermen were the only significant participants.[3] The fishing crews of both nations had established themselves

---

[1]This essay appeared originally in *The Northern Mariner/Le Marin du nord*, XI, No. 1 (2001), 1-10.

[2]See John Mannion and Selma Barkham, "The 16th Century Fishery," in R.C. Harris (ed.), *Historical Atlas of Canada, I: From the Beginning to 1800* (Toronto, 1988), plate 22 (*HAC*).

[3]The French fishery was roughly twice as large as its English counterpart. At its peak between 1678 and 1688, the French fishery employed 10,000 to 12,000 men (about one-quarter of the French maritime population) and over 400 vessels, producing more than 500,000 quintals annually during the 1680s, a time when the British fishery scarcely produced a fifth that amount; Laurier Turgeon, "Colbert et la pêche française à Terre-Neuve," in Roland Mousnier (ed.), *Un Nouveau Colbert: Actes du Colloque pour le tricentenaire de la mort du Colbert* (Paris, 1985), 258; and Keith Matthews, "A

in fairly specific regions by then as well. Thus, the English dominated the Avalon Peninsula from Cape Race to Cape Bonavista (the so-called "English Shore"); French fishermen predominated along the south coast, with a fortified settlement at Plaisance. There were also important installations on the island of St. Pierre, as well as the "Petit Nord," as the region between Cape St. John and Quirpon was known. For their part, fishermen from the Basque region of southwestern France made the west coast of the island their special preserve.[4]

The precise limits of these fishing zones had been established by custom long before emerging state machinery became involved. Indeed, throughout the seventeenth century, the English government adopted the position that the North American fisheries should be encouraged but not rigorously administered by the state. Though France established an outpost at Plaisance after 1660 to support the fishery, no attempt was made to extend the effective range of French authority beyond Placentia Bay.[5] This only began to change in the eighteenth century as attitudes to the "state" itself took firmer shape. Under the influence of mercantilism both England and France began to regard the fisheries as strategic and economic assets of the first order. The strategic importance stemmed in part from the perception of the fishery as a "nursery for seamen" which not only employed thousands of landsmen every year but also transformed them into the kind of experienced mariners prized by navies in the event of war. In terms of direct employment, consumption of domestic goods and services, and the generation of a favourable balance of trade with other mercantile powers, the fishery was a source of great national wealth and, according to mercantilist logic, of great national power. As a result, the fishery was so highly prized by both countries that neither would willingly give it up, either in whole or in part. During attempts in 1761 to negotiate an end to the Seven Years' War, members of the British and French governments independ-

---

History of the West of England–Newfoundland Fisheries" (Unpublished DPhil thesis, Oxford University, 1968), 158-162 and 191.

[4]For the English Shore, see Matthews, "History," 187; for the French domain, see John Mannion and Gordon Handcock, "The 17th Century Fishery," in *HAC*, I, plate 23.

[5]A succinct discussion of British policy appears in Sean Cadigan, *Hope and Deception in Conception Bay: Merchant-Settler Relations in Newfoundland, 1785-1855* (Toronto, 1995), 27-28. The first British parliamentary legislation to concern itself with the fishery at Newfoundland (10 and 11 Wm. III c. 25) was passed in 1699. Though France established a colony at Plaisance in 1660, its navy did not escort or supervise French fishing ships or patrol the "Petit Nord." John Humphreys, *Plaisance: Problems of Settlement at this Newfoundland Outpost of New France, 1660-1690* (Ottawa, 1970); and Jean-François Brière, "The Safety of Navigation in the French Codfishing Industry at Newfoundland in the 18th Century" (Unpublished paper presented at the annual meeting of the Canadian Nautical Research Society, Galiano Island, BC, July 1986).

ently ventured the same opinion: that the Newfoundland fishery was more valuable than Canada and Louisiana combined "as a means of wealth and power."[6]

Notwithstanding the importance that Europeans attached to the fisheries by the late 1600s, there were surprisingly few charts of the island and almost none of particular bays or stretches of the coast. The French prepared some cartographic impressions of the island, giving predictable attention to Plaisance and Placentia Bay. There were, however, very few British charts. One noteworthy exception was the map compiled by Captain John Mason and published in 1624 - the first deliberate attempt to survey the Avalon Peninsula. Another was Southwood's map of 1677, published in John Thornton's *English Pilot* in 1689.[7] Nevertheless, at best these were crude approximations of the island and of little use to navigation. In considerable measure, this reflected not disinterest but the existing limitations of scientific cartography. Indeed, it was not until 1675 that two noteworthy events - the founding of the Royal Observatory at Greenwich and the publication of the first English sea atlas, John Sellers' *Atlas Maritimus* - signalled that determined efforts were being made to improve the standards and extent of English knowledge of the oceans.[8] France, too, took steps to promote the quality and reliability of navigation and cartography, establishing a hydrographic office in 1720 even as navigation was being taught on both sides of the Atlantic. The immense hazards to navigation in the Gulf and River of St. Lawrence (likened by one early eighteenth-century traveller to "walking blind-folded and barefoot in a room of jagged glass")

---

[6]The Board of Trade (BT) maintained that "the Newfoundland Fishery as a means of wealth and power is of more worth than both of the aforementioned provinces;" Gerald Graham, "Fisheries and Sea Power," Canadian Historical Association, *Annual Report 1941* (Ottawa, 1942), reprinted in G.A. Rawlyk (ed.), *Historical Essays on the Atlantic Provinces* (Ottawa, 1967), 8. At approximately the same time, the French Minister of Marine, the Duc de Choiseul, insisted that "the codfishery in the Gulf of St. Lawrence is worth infinitely more for the realm of France than Canada or Louisiana." Cited in Max Savelle, *The Origins of American Diplomacy: The International History of Anglo-America, 1492-1783* (New York, 1967), 475n. See also Jean-François Brière, "Pêche et politique à Terre-Neuve au XVIIIe siècle; la France véritable gagnante du traité d'Utrecht?" *Canadian Historical Review*, LXIV, No. 2 (1983), 168-170.

[7]"Cartography," *Encyclopedia of Newfoundland and Labrador* (5 vols., St. John's, 1981-1994), I, 372 (*ENL*); J.C. Beaglehole, *The Life of Captain James Cook* (Stanford, 1974), 63; and Rodney W. Shirley, "The Maritime Maps and Atlases of Seller, Thornton, and Mount and Page," *Map Collector*, No. 73 (1995), 2-9.

[8]Ian K. Steele, *The English Atlantic 1675-1740: An Exploration of Communication and Community* (Oxford, 1986), 15.

even inspired a major cartographic effort there in the 1730s.[9] Nevertheless, the output of the French hydrographic office was not impressive before mid-century. Meanwhile, the British did not even appoint a Hydrographer of the Navy until 1795, and the single greatest challenge to reliable oceanic navigation, namely the accurate measurement of longitude, would not be resolved until the last quarter of the eighteenth century.[10] And while shipmasters were increasingly expected to have some familiarity with "scientific navigation," the fact remained that a master's chief guides were still, in the words of Ralph Davis, "dead reckoning with compass and log."[11]

Small wonder, then, that cartography seemingly played so small a role in the voyages to Newfoundland. Charts did exist, but these were of lesser significance than familiarity with prevailing winds and ocean currents, the clues revealed by a sounding lead and the knowledge acquired from those who had sailed a route before. Years of accumulated experience, earned through lengthy apprenticeship, not charts, governed the movement of merchant shipping at Newfoundland well into the eighteenth century.[12] Thus, in 1726, as soon as flocks of sea birds and pods of whales made their appearance during the voyage to St. John's of the Scottish merchantman *Christian*, the crew relied on the sounding lead to guide the ship to its landfall, despite the master's demonstrable ability to measure positions on a chart.[13] Nor was the master

---

[9]James Pritchard, *Louis XV's Navy, 1748-1762: A Study of Organization and Administration* (Montréal, 1987), 21; and Ken Banks, "'Lente et assez fâcheuse Traversée:' Navigation and the Transatlantic French Empire, 1713-1763," in A.J.B. Johnston (ed.), French Colonial Historical Society, *Proceedings* (Cleveland, 1996), 85.

[10]Pritchard, *Louis XV's Navy*, 21; and G.S. Ritchie, *British Naval Hydrography in the Nineteenth Century* (London, 1967; reprint, Edinburgh, 1995), 30.

[11]Ralph Davis, *The Rise of the English Shipping Industry in the Seventeenth and Eighteenth Centuries* (London, 1962; reprint, St. John's, 2012), 116-118.

[12]*Ibid.*, 109 and 116-118.

[13]Scottish Record Office, RH 9/14/102, "Journal of a Voyage to Newfoundland, in 1726-1727 kept by Edward Burd, Jr., supercargo." Burd recorded measurements of latitude and longitude following the ship's departure from Spain in 1727, but not during the earlier leg of the voyage from Scotland to Newfoundland; see daily entries throughout July 1726, and entries for March and April 1727. In all likelihood, the master calculated the ship's position at every stage of the voyage. Commenting on the challenge of navigating in Newfoundland waters, a passenger on the French *flûte Eléphant* observed in 1729 that "When, as is common in these waters, navigators are unable to take altitudes because of fog or bad weather, they are compelled to go under bare poles, and frequently even have to heave aback during the night; otherwise they would risk becoming stranded on these shores;" cited in Gilles Proulx, *Between France and New France: Life aboard the Tall Sailing Ships* (Toronto, 1984), 64. A most dra-

relying on his own experience, for he had never before been to Newfoundland. Rather, prior to departure he acquired the foreknowledge of the thousands of British mariners, passed on by word of mouth or in "waggoners" (sea charts valued more for their annotations than for their cartographic accuracy).[14] Only when such accumulated experience was lacking, when prior knowledge of a coast or its resources did not exist, did the mariner hesitate to proceed.

This was the situation facing British venturers in the Newfoundland fishery after 1713. Under the terms of the Treaty of Utrecht, the south side of the island, including the islands of St. Pierre and Miquelon, became British territory. France retained only fishing privileges on the so-called "French Shore," which extended north from Cape Bonavista to Point Riche on Newfoundland's west coast.[15] All French civil and military authorities at Plaisance, together with most of the inhabitants, were therefore evacuated by 1714.[16] Thus, in one fell swoop the British had greatly extended the theoretical territorial limits of their fishery. But while the authorities were convinced that the region was rich in fish and other desirable resources, everything to the west of Trepassey was *terra incognita* to British fishermen. Moreover, British authorities were cautioned that French merchants would not give up their lucrative investments in Newfoundland quite so easily but would continue to trade clandestinely until British commerce pushed into the region.[17] In a letter to Secre-

---

matic illustration of the degree to which navigation in Newfoundland waters depended more on experience than science is provided by the experience of the French fleet under the Duc d'Enville in 1746 as it groped its way into North American waters with inadequate charts and too few pilots; see James Pritchard, *Anatomy of a Naval Disaster: The 1746 French Expedition to North America* (Montréal, 1995), esp. chaps. 4-6.

[14]Davis, *Rise of the English Shipping Industry*, 123-124; and Shirley, "Maritime Maps," 2.

[15]James K. Hiller, "The Newfoundland Fisheries Issue in Anglo-French Treaties, 1713-1904," *Journal of Imperial and Commonwealth History*, XXIV, No. 1 (1996): 1-23; and Brière, "Pêche et politique à Terre-Neuve," 168-187.

[16]Jean-Pierre Proulx, *Placentia, Newfoundland* (Ottawa, 1979), 51-52 and 118-119. In 1713, the civilian population of French Newfoundland numbered about two hundred, a figure which would double were one to include the contract fishermen; Humphreys, *Plaisance*, 6. A small number of inhabitants were prepared to swear the English oath of allegiance and remain in Newfoundland. Proulx estimates that fifty to sixty people remained behind at Placentia after the evacuation. There were others in St. Pierre, the Burin Peninsula and Fortune and Hermitage bays.

[17]Great Britain, National Archives (TNA/PRO), Colonial Office (CO) 194/5, 117-117v, Moses Jacqueau to William Lownde, Treasury, 7 May 1714. Jacqueau predicted that French merchants would use English prize ships and hire English masters to front the continuation of their trade with Newfoundland.

tary of State Lord Stanhope, the Board of Trade (BT) concluded that "it is necessary there be a Survey made of the late ffrench part of Newfoundland, for that thereby many good harbours & ffishing Places may be discover'd wch will incourage our Fishing Ships to resort thither, who are now unacquainted with that Coast."[18] In fact, British authorities commissioned not one but two surveys of Newfoundland's south coast immediately following the Treaty of Utrecht.

The first of these was undertaken by William Taverner, who was appointed by the BT in 1713 to conduct a survey of the coast west of Placentia. Taverner was a Newfoundland planter, merchant and shipowner who divided his time between Poole and his fishing operations in Trinity Bay.[19] He had connections with the London merchant community and had been involved on its behalf in developing a cod fishery on the northwest coast of Scotland during the later years of the War of the Spanish Succession. He also engaged in privateering against the more remote French outposts in Newfoundland during that war.[20] By 1712, he was submitting memoranda to the BT on French possessions in Newfoundland. Whether he had any training as a cartographer or surveyor is not known. What we can confidently assume he had in abundance was familiarity with the general conditions and specific hazards of sailing on the Newfoundland coast. Taverner was a seasoned shipmaster, undoubtedly familiar with the rudiments of chart making. More important, he had a practical understanding of navigation in those waters. As such, he was able to establish himself in the eyes of both merchants and officials in London as an authority on Newfoundland, on the cod fishery and on the coastal region which the French began to vacate in 1713. To them, he was an obvious choice to undertake the survey so desired by the BT.

---

[18]TNA/PRO, CO 5, vol. 4i, fol. 28, BT to Secretary of State Stanhope, 10 March 1715.

[19]David B. Quinn, "William Taverner," in George Brown, David Hayne and Frances G. Halpenny (gen. eds.), *Dictionary of Canadian Biography, Vol. 3: 1741-1770* (Toronto, 1974), 617-619 (*DCB*); Matthews, "History," 323; and Robert Cuff, "Taverner's Second Survey: Introduction," *Newfoundland Quarterly*, LXXXIX, No. 3 (1995), 9-10. Gordon Handcock, *Soe longe as there comes noe women: Origins of English Settlement in Newfoundland* (St. John's, 1989), 47, refers to the Taverners as "one of the more remarkable pioneer families to have established themselves in Newfoundland."

[20]TNA/PRO, CO 194/5, 109, Petition of William Taverner to Lord Oxford, Lord High Treasurer, n.d., enclosed with Taverner to the Lords Commissioners for Trade and Plantations, 31 March 1714. Taverner expands on his privateering in Fortune Bay in TNA/PRO, CO 194/6, 231v, William Taverner, "Second Report...of His Survey Work, 1714/1715."

Taverner's instructions were quite clear: he was to survey the region and to take stock of its resources. Though Taverner later indicated that he made some maps, nothing in his instructions specifically required him to do so; the emphasis was on gathering information.[21] It was equally plain that his work was also intended to signal the assertion of British sovereignty in a region still being evacuated by the French.[22] Taverner was therefore to take a census and administer an oath of allegiance to any French inhabitants who wished to remain in Newfoundland. Acting with commendable energy, he arrived at Placentia in late June 1714, but remained only long enough to prepare the *Tyger* galley, a vessel which John Moody, the garrison commander, had put to his use. On 27 July he set out for the island of St. Pierre (St. Peter's to the English), Bay d'Espoir and Hermitage Bay, returning to Placentia in September. It is clear that this was a fairly superficial reconnoitre of the region – he appears to have by-passed the inner recesses of Fortune Bay completely, and even in Hermitage Bay, he only visited Grole and "Isle Espere" (Pass Island). The time available to him simply did not permit a more extensive exploration of the south coast. Though he prepared a detailed report on this initial survey for the BT, including a chart of "the islands and harbour of St. Peter's, with the island of Columba and the adjacent rocks," most of the anecdotal details were based on what he already knew before the survey began.[23]

Taverner next prepared to undertake a survey that would concentrate on Placentia Bay and the Burin Peninsula. Having determined by experience that the *Tyger* galley "sail'd badly" and that "a Shallop was absolutely Necessary for the Surveying of a great many Places," Taverner discharged *Tyger* and hired *Delore*, which he despatched to the tiny community of Burin to serve as a supply depot. He then set out in late October from Placentia in a shallop, working his way counter-clockwise around Placentia Bay until he met with his supply vessel at Little Burin on 12 December 1714. The onset of winter forced him to suspend any further work until the following year.

---

[21]Taverner's instructions and additional instructions of 21 and 22 July 1713 may be found in TNA/PRO, CO 194/5, 99-103v.

[22]The ship *Héros*, with Philippe Pastour de Costebelle, governor of the former French colony there since 1695, did not depart Placentia for Cape Breton Island until 25 September 1714; TNA/PRO, CO 195/6, 25-28, Col. John Moody to BT, 9 September 1714 (OS); and Georges Cerbelaud Salagnac, "Philippe Pastour de Costabelle," *DCB*, I, 509-513.

[23]TNA/PRO, CO 194/5, 260-262, Taverner to Secretary of State, 22 October 1714. See also Quinn, "Taverner," 618; Cuff, "Taverner's Second Survey: Introduction," 10; and C. Grant Head, *Eighteenth Century Newfoundland: A Geographer's Perspective* (Ottawa, 1976), 57-58.

By March Taverner was clearly impatient to resume his survey. With *Delore* still frozen in, he took the shallop on a brief foray around the Burin Peninsula as far as Grand Bank before rejoining his supply ship. It was at this point that the survey suffered three setbacks that put an end to Taverner's efforts. First, his boat was swamped on the rocky coast. Valuable supplies and equipment were lost, and his men, soaked by the icy waters, suffered from exposure, and one man endured frostbite. A worse disaster was the loss of *Delore* just as it was released from the ice and was about to sail west in support of further surveying.[24] The most serious setback, however, came when Taverner made his way back to Placentia for further orders and, he hoped, another vessel, only to discover that Colonel Moody had complained to the senior naval officer there that Taverner had "wholly left off his Surveying" and was spending most of his time at St. Pierre "where he applys himself to fishing and merchandizing."[25] Taverner was ordered to discontinue his work and to return immediately to London where he was to account for his activities.

In London Taverner appears to have satisfied the BT of the worth of his work even though he had only visited some of the areas he described. His reports stressed the region's abundant natural resources – the location and size of beaches which until the evacuation had supported French fishing stations, their proximity to productive fishing grounds, and the ubiquity and quality of the timber – so suitable for "sparrs, Masts for Ships, very good Timber of all Sorts both for Boards and Plank" – and the plentiful game. Taverner clearly was impressed with the potential of the area. He declared "the very worst" of the fishing grounds and shore stations to be "better than our former English Settlements to the NEward," claiming that "one boat at St. Peters &ca. have taken as much fish...as Three boats at most of the former English Settlements."[26] Indeed, the island of St. Pierre, in his "humble Opinion...exceeds all the rest for Codfishing, its a good harbour, and beech might be made for 300 Boats." Yet Taverner also emphasized the dangers of this new and still unfamiliar coast, dangers which his own experiences had amply demonstrated.[27] Perhaps even more alarming were his claims that French ships making their way to Cape Breton Island had stopped at St. Pierre and engaged in trade; that

---

[24]Taverner, "Second Report," 226-241.

[25]TNA/PRO, Admiralty (ADM) 1, vol. 1778, IX, Capt. Edward Falkingham to Moody, and Moody to Falkingham, 17 August 1715.

[26]TNA/PRO, CO 194/5, 262, Taverner to Secretary of State, 22 October 1714; and Taverner, "Second Report."

[27]"That part of Newfoundland which I have Survey'd is very Dangerous, in many places as appears by my Chart." Taverner, "Second Report."

the handful of French fishermen who had sworn an oath of allegiance to the British crown in order to remain in Newfoundland had participated in that trade; and that people from the new French colony on Cape Breton Island were coming to Newfoundland during the winter to hunt and to trap for furs.

These were disturbing allegations for, if true, they suggested that British control over the newly acquired region was uncomfortably tenuous, perhaps even in jeopardy. The obvious solution to French encroachment on so valuable a territory was for British fishermen to push into the region in substantial numbers. This, however, had not yet happened, nor was it likely to do so in the foreseeable future, in Taverner's opinion. The region was simply not sufficiently well charted, he said, "wch at present deterrs the English from sending their Ships to Fish and Trade there."[28] Taverner had, of course, produced a couple of charts which, in the later opinion of the BT, "have been of great use."[29] And he could not resist reminding the Board that had he been provided with another vessel and proper support after his return to Placentia "in all Probability, I might have brought home with me an Exact Chart from Cape Race to Cape Les Anguiles, or the Isles of St. George."[30] Yet Taverner's skills as a cartographer were questioned by some, and the value of his survey appears to have rested more with his resource inventory and commentary than with the quality of his charts.[31] There were also the accusations of Colonel Moody and others. British authorities therefore decided that the task of preparing a proper cartographic survey of the south coast would go to someone else. That person was John Gaudy, who served as a midshipman extra on *Worcester* in 1715 when that warship was stationed in Newfoundland. *Worcester's* captain and the commander of the Newfoundland station that year was Thomas Kempthorne, and it was he who brought Gaudy to the attention of his superiors for the young man's chart-making skills.[32]

---

[28]TNA/PRO, CO 194/5, 262, Taverner to Secretary of State, 22 October 1714.

[29]PRO, CO 195/7, 145, BT to Duke of Newcastle, 30 March 1727. The only example of Taverner's cartographic work to survive would appear to be the map of St. Pierre published in Richard Mount, William Mount and Thomas Page, *English Pilot* (4 vols., London, 1716); see "Cartography," *ENL*, I, 373.

[30]Taverner, "Second Report."

[31]Some British merchants engaged in the Newfoundland fishery questioned Taverner's qualifications as surveyor; Quinn, "Taverner," 618.

[32]TNA/PRO, ADM 1/2006, I, Letters of Capt. Thomas Kempthorne (Kempthorne Letters), esp. Kempthorne (*Worcester*, St. John's) to Secretary to the Admiralty Josiah Burchett, 6 October 1715. In the aftermath of the War of the Spanish Succession there were far more lieutenants than suitable vacancies. A midshipman extra was some-

Kempthorne was an energetic officer who took his responsibilities in Newfoundland quite seriously. Thus, in contrast to most others who commanded the Newfoundland station during the first half of the eighteenth century, his reports included exhaustive and perceptive observations, criticisms and recommendations on the administration of the fishery.[33] Kempthorne had arrived at his own conclusions concerning the need to learn more about the former French regions of Newfoundland, for he had already given Gaudy "an opportunity of draughting the SEt part of [Newfoundland], and the harbours thereto belonging." Stressing that the region "has been by others very negligently & erroniously discribed," Kempthorne was convinced that "a true description" of that coast "is very much wanted by our Trade." He sent Gaudy back to London with the results of his surveying and a recommendation to the Admiralty that he be appointed to conduct a more extensive survey of the major fishing harbours on the south coast in the following year, adding that "there can't be recomended to their Lordships a person whose ability, and meritt, can better qualify him for that performance: and I doubt not but Sr Chals Wager will speake as large in his character."[34]

Whether it was because of Kempthorne's endorsement, his reference to the newly appointed Controller of the Navy or perhaps simply the quality of Gaudy's chart of "the Sea Coast of Newfoundland from ye Bay of Bulls to Little Placentia," the recommendation was endorsed by the Admiralty. In the full conviction that "the Survey of these Coasts is of Consequence to the Service," Gaudy was appointed a lieutenant on HM Sloop *Swift* under Captain Thomas Durell in June 1716 and ordered to Newfoundland to carry out a cartographic survey "of its Coast and harbours thereto belonging."[35] To this end,

---

one who held the rank of lieutenant but who shipped aboard a warship in the billet of midshipman hoping that a lieutenancy would fall vacant; see Daniel Baugh (ed.), *Naval Administration, 1715-1750* (London, 1977), 37.

[33]Kempthorne Letters, esp. his undated 1715 letter to Burchett. A copy can be found in TNA/PRO, CO 194/5, 379-386v.

[34]Kempthorne Letters, Kempthorne to Burchett, 6 October 1715.

[35]TNA/PRO, ADM 2/449, Burchett to Capt. Thomas Durell, *Swift* sloop, 13 June 1716; National Maritime Museum (NMM), ADM L/S 587, "Journal Kept by John Gaudy, Lieutenant of his Majtys Sloop ye Swift (17 June 1716-12 September 1717)" (Gaudy, "Log"). The chart Gaudy had prepared for Kempthorne, and by which he now brought himself to the attention of the Admiralty, was subsequently published; see Mount and Page, *English Pilot*, IV, "A Chart shewing Part of the Sea Coast of Newfoundland from ye Bay of Bulls to Little Placentia, exactly and carefully lay'd down by Iohn Gaudy, Anno 1715," which includes an inset map of Trepassey harbour. A copy of Gaudy's "Draught and description of Plecentia [sic] Harbour" is in the Naval Historical Library in London.

Swift had been supplied with an extra small boat, twelve to fourteen feet in length, an extra anchor and cable and enough old canvas to make an awning for the full length of the ship. As well, a cabin had been built for Mr. Gaudy "in the most convenient place for his purpose in the Steeridge of the said Sloop." Meanwhile, Gaudy visited the shop of John Bollinger to purchase the surveying equipment he required: a theodolite, a plane table, a brass ruler and sights, a quadrant, a Gunter's chain with a large pair of compasses, a sector scale and a protractor "with 2 pair compasses in a case," a large sextant with a telescope and a rack and three-log staff. Of equal importance was a visit to the premises of Richard Mount and Company where Gaudy purchased the paper his surveying and chart-making would require. Finally, the Admiralty authorized a disbursement of £100, not only to pay for this equipment but also "for subsisting his Family in his Absence."[36] It must have been a welcome reversal of fortune for a man who one year earlier had faced an uncertain future as an unemployed naval lieutenant.

Swift got underway in late June and by early August made landfall at Cape Broyle, just south of St. John's. There was a brief delay in Trepassey harbour while the ship replenished its water and wood and brewed a supply of spruce beer. The sloop-of-war then proceeded to St. Pierre, where Gaudy transferred to its yawl to begin his work.[37] For nearly a week, Gaudy was engaged in a survey of the former French island and its harbour before *Swift* got underway once again and proceeded west as far as Ramea and Burgeo. Another week or so was spent surveying the coast in that neighbourhood before *Swift* began working its way east. Stopping for still more surveying work at Bay d'Espoir and Fortune Bay, Gaudy returned to St. Pierre and then went on into Placentia Bay, following the coast of the Burin Peninsula, before arriving at Placentia on 2 October, where Gaudy's survey came to an end.

The two months since *Swift* arrived in Newfoundland had not passed uneventfully. For instance, there had been the incident involving the ship's carpenter, "a very fractious fellow & Mutinous Man" in Gaudy's masterfully understated opinion. During their visit to Trepassey, the carpenter became drunk and abusive, going so far as to threaten Captain Durell with a broad-

---

[36]NMM, ADM/A/2057B, Admiralty Out-Letters to the Navy Board, 9 June 1716. According to Ritchie, *Admiralty Chart*, 25, a plane table was "a simple form of portable plotting board which could be set up on a tripod at various stations and upon which conspicuous points could be fixed by plotting intersecting bearings."

[37]Gaudy, "Log." See also TNA/PRO, ADM 51/961, IV, "Journal keept [*sic*] on board his Majtys Sloop the Swift by me Thos Durrell Commandr" (17 May 1716-23 December 1718); and Durrell, "Log."

sword.[38] Later, *Swift* captured a boat-load of soldiers who had deserted from the garrison at Placentia and were making for Cape Breton Island when they were driven ashore on St. Pierre.[39] And from time to time, the crew was allowed ashore to hunt for caribou and geese. For the most part, however, Gaudy's work was routine, even tedious. It was rendered all the more so when (as was frequently the case) the weather turned "very foggy, Raining & blowing" and slowed his work. Still, Durell (and presumably Gaudy) were pleased by the summer's labours. As Durell explained to Admiralty Secretary Josiah Burchett, the first month's efforts alone had resulted in "a Compleat Draft of this Coast from [Burgeo] to [St. Pierre], with the Harbours & Islands wch are many fitt for the fishery, & not taken Notice of in any Charts before, nor Known by our Traders."[40]

　　　Whether Gaudy's survey – or Taverner's, for that matter – had the desired effect is difficult to determine. Placentia would eventually develop in the 1730s into the centre of a thriving migratory ship fishery based in the ports of North Devon. But throughout the 1720s, British fishermen remained reluctant to push into the region, so that government continued to express concern at the rate of growth.[41] And notwithstanding continued reports that French fishermen persisted in visiting the south coast, and that French merchants were trading in the region, the British lacked the resources to patrol the coast west of St. Pierre and the Burin Peninsula.[42] The surveys of William Taverner and John Gaudy were both intended to open Placentia Bay and the south coast to British exploitation, and, by promoting a British presence there, to assert British sovereignty. Instead, both Taverner's and Gaudy's charts appear to have vanished. Not until James Cook's cartographic survey of the south coast would detailed charts of the region be made widely available to mariners.

　　　Until then, Taverner and Gaudy exemplified the two major influences guiding early eighteenth-century navigation and cartography. Taverner was the embodiment of the practical sailing master. Experienced with the characteristi-

---

[38]Gaudy, "Log," 12 August 1716. Following completion of the survey, while *Swift* was moored in Placentia harbour, the carpenter became drunk and quarrelsome again and began "beating...with a Red hot polker." *Ibid.*, 13 October 1716.

[39]TNA/PRO, ADM 1/1694, No. 9, letters of Capt. Thomas Durell, Durell (*Swift*, St. Peter's) to Burchett, 16 September 1716.

[40]*Ibid.*

[41]Head, *Eighteenth Century Newfoundland*, 58-59.

[42]See, for instance, Falkingham's complaints in 1732 in TNA/PRO, CO 194/9, 212-216v, "Answers to the Heads of Inquiry;" and PRO, ADM 1/1779, Falkingham to Burchett, 4 October 1732.

cally difficult conditions in Newfoundland waters, he relied more on instinct and accumulated judgment than on instruments and mathematical measurement. Indeed, the only navigational device mentioned in his reports was an azimuth compass, lost when his shallop was swamped in April 1715. In contrast, Gaudy was a naval officer, junior in rank if not in years, and an accomplished cartographer. Skilled with the instruments of his day, he knew how to translate their measurements into accurate charts. That both men were employed to survey the same stretch of coast within a year of each other should not be perceived as evidence of bureaucratic redundancy. Rather, this points to two not entirely unrelated early eighteenth-century developments. First, we catch a glimpse of cartographic surveying in which two kinds of talents were highly valued, one practical and the other scientific. Second, these early efforts to survey the Newfoundland coast underscore the growing interest of mercantilist British officials in asserting sovereignty in a region deemed likely to have important commercial and strategic significance for the future.

The Problem of Piracy in the Newfoundland Fishery in the Aftermath of the War of the Spanish Succession[1]

In his classic *History of Newfoundland*, written at the close of the last century, Daniel Woodley Prowse dismissed the period from 1714 to 1727 as akin to "the heavy prosaic Hanoverian monarch – dull, uneventful, peaceful, and prosperous."[2] In fact, we now know that it was a period of profound changes and hardships in the fishery, frequently tumultuous and punctuated occasionally by moments of violence and terror. It was a period when the inshore fishery failed, causing such widespread destitution among the permanent residents that one observer was led to remark that they were "worse off than negroes and slaves."[3] It was an age when British policy was guided by the firm conviction that Newfoundland was a seasonal fishing station which required no government, no courts, no constabulary, not even a militia. There were, in short, no institutions of law or order to contain the social tensions that ensued from the conditions that prevailed in Newfoundland after 1715. The owners of property in St. John's were sufficiently disturbed by what they perceived as the threat of lawlessness and anarchy that they established a court of law late in 1723. Conscious of the fact that they had no authority from England to do so – indeed, conscious that they were violating existing law practices and laws – they justified their initiative with reference to John Locke's *Essay Concerning the True Original, Extant, and End of Civil Government* (1690).[4] It was also largely in consequence of the failure of the inshore fishery that, for the first

---

[1]This essay appeared originally in Poul Holm and Olaf Janzen (eds.), *Northern Seas Yearbook 1997* (Esbjerg, 1998), 57-75.

[2]Daniel Prowse, *A History of Newfoundland from the English Colonial and Foreign Records* (London, 1895; reprint, Belleville, ON, 1972), 283.

[3]Great Britain, *Calendar of State Papers Colonial (CSPC)*, America and the West Indies, 1714-1715, No. 179 (i), James Smith to Townshend, 1715, cited in Keith Matthews, "A History of the West of England-Newfoundland Fisheries" (Unpublished DPhil thesis, Oxford University, 1968), 314. According to Matthews, Smith had been sent to Newfoundland in 1710 to act as judge of a Vice-Admiralty court that was never put into action.

[4]Jeff A. Webb, "Leaving the State of Nature: A Locke-Inspired Political Community in St. John's, Newfoundland, 1723," *Acadiensis*, XXI, No. 1 (1991), 156-165; and Matthews, "History," 355-356.

time ever, the English began to send their fishing crews offshore to the great fishing banks.[5] Finally, it was in the midst of all these profound stresses and responses that piracy returned – briefly but sharply – to Newfoundland. This paper seeks to explain the appearance of pirates in Newfoundland between 1717 and 1725. It will do so by placing developments in Newfoundland within the context of the general resurgence of piracy which occurred throughout the Anglo-American North Atlantic world during the decade after the conclusion of the War of the Spanish Succession. At the same time, it will link the appearance of pirates in Newfoundland to the particular needs of early eighteenth-century pirates as well as to the conditions that prevailed in the fishery after 1713.

 Falconer's Marine Dictionary defined "pirate" as "a sea-robber, or an armed ship that roams the seas without any legal commission, and seizes or plunders every vessel she meets indiscriminately, whether friends or enemies."[6] It had not always been that simple. During the sixteenth century, piracy was often an activity officially sanctioned by emerging Western European powers like England, France and Holland as they struggled to capture a share of the wealth pouring out of the New World on Spanish and Portuguese shipping. The complex entanglement of wars that prevailed for much of the second half of that century helped to blur the tenuous distinction between legitimate commerce-raiding against declared enemies and piracy. One nation's seafaring hero was another's villainous pirate.[7] Only as officially sanctioned piracy faded in the seventeenth century, as Spain's rivals became imperial powers themselves, and maritime empires – and commerce – expanded, did European piracy conform more consistently to Falconer's definition.[8] Improvements in oceanic shipping, which encouraged not only trade but also a shift to deep-sea

---

 [5]Matthews, "History," 306-307; and C. Grant Head, *Eighteenth Century Newfoundland; A Geographer's Perspective* (Ottawa, 1976), 63-65 and 72-73.

 [6]William Falconer, *Falconer's Marine Dictionary* (London, 1780; reprint, Paderborn, 2012), 215.

 [7]Robert C. Ritchie, *Captain Kidd and the War against the Pirates* (Cambridge, MA, 1986), 10; K.R. Andrews, "The Expansion of English Privateering and Piracy in the Atlantic, c. 1540-1625" and David Quinn, "Privateering: The North American Dimension (to 1625)," both papers presented at a colloquium on "Piracy and Privateering" sponsored by the International Commission for Maritime History, San Francisco, 1975, and subsequently printed in an unpublished collection of colloquium *Proceedings*. Francis Drake, one of Elizabethan England's most famous heroes, earned his knighthood for the success with which he attacked Spanish shipping during his circumnavigation of the world in 1577-1580. Yet Drake's exploits were piratical because England and Spain were then at peace.

 [8]Ritchie, *War against the Pirates*, 16.

piracy, the inability of national navies to control or police trade and the conclusion in 1603 of England's war with Spain, which also meant an end to legitimate commerce raiding, all contributed to an increase in piracy during the opening decades of the seventeenth century, particularly in European waters.[9]

Neither the fishery nor the fish trade from Newfoundland was exempt from piracy during this early period. In the closing decades of the sixteenth century, piracy at Newfoundland – particularly of the "officially sanctioned" type – was "endemic" in the words of one authority.[10] Portuguese and Basque fishing fleets suffered irreparable harm from piratical attacks.[11] The collapse in turn of the Iberian fisheries at Newfoundland stimulated the expansion of English and French fisheries at Newfoundland as the Iberian markets turned to foreign suppliers for saltfish. It was therefore primarily English and French merchantmen delivering fish to the southern European markets who fell prey to the most feared pirates of the first half of the seventeenth century, the North African raiders known as "Sallee Rovers." Based in the ports of modern-day Morocco, Tunisia and Algeria, they were a constant menace to the Mediterra-

---

[9]*Ibid.*, 17. See also Andrews, "Expansion," 217-218, as well as Clive Senior, "The Confederation of Deep-Sea Pirates: English Pirates in the Atlantic 1603-25," which was also presented at the 1975 San Francisco colloquium on "Piracy and Privateering." As well, see C. L'Estrange Ewen, "Organized Piracy Round England in the Sixteenth Century," *Mariner's Mirror*, XXXV, No. 1 (1949), 29-42; and John C. Appleby, "A Nursery of Pirates: The English Pirate Community in Ireland in the Early Seventeenth Century," *International Journal of Maritime History*, II, No. 1 (1990), 1-27.

[10]"Privateering and Piracy Become Endemic, 1584-1596," in David B. Quinn (ed.), *New American World: A Documentary History of North America to 1612, Vol. IV: Newfoundland from Fishery to Colony, Northwest Passage Searches* (New York, 1979). The English privateer Hugh Jones was forced to stop at Newfoundland in 1584 after a successful cruise against Spanish trade in the Caribbean; when disease decimated his ship's crew. French Basques, who had frequently been victimized by the mounting hostility between England and Spain, seized Jones' ship and plundered it – technically, an act of piracy; Quinn (ed.), *New American World*, IV, 45-46. A similar incident involving another English ship occurred in 1590; Quinn (ed.), *New American World*, IV, 53-54.

[11]Bernard Drake set out for Newfoundland in 1585 at the head of a squadron of ships to warn English merchants that Spain had clapped an embargo on English shipping in Spain and to attack Spanish shipping in the fishery. In fact, Drake avoided the area where the Spanish fished in the greatest numbers, cruising primarily against Portuguese fishing ships – sixteen or seventeen in total – and whatever merchantmen had the misfortune to stumble across his path; David B. Quinn, "Bernard Drake," in George W. Brown (gen. ed.), *Dictionary of Canadian Biography, Vol. I: 1000-1700* (Toronto, 1966), 278-280 (*DCB*).

nean trade. Indeed, by 1611, they had extended their cruising range north as far as the English Channel itself.[12]

Nor were the fishing fleets at Newfoundland necessarily any safer; though the Sallee Rovers rarely if ever roamed there, a number of European pirates cruised in Newfoundland waters.[13] Of these, the most notorious was undoubtedly Peter Easton, who began his career attacking French ships in the English Channel in 1610 before crossing the Atlantic in 1612 where, with six ships under his command, he plundered not only English, French and Portuguese ships at Newfoundland but interfered as well with the first attempt to colonize the island at Cupid's Cove in Conception Bay.[14] According to Clive Senior, the attraction of Newfoundland for seventeenth-century pirates lay not in plunder, since "the seizure of boat-loads of fish was not in itself a particularly attractive prospect for any self-respecting pirate," but in the provisions, equipment and other supplies they could steal and the men they could conscript into their own crews.[15] Easton's raid certainly bore testimony to this. According to witnesses, Easton arrived at Newfoundland after cruising on "the Coast of Guinnie" and promptly began to extort food, sails, rigging and men from the fishing ships operating there.[16] All accounts agree that the Caribbean, particularly the Spanish trade routes between America and Europe, were the preferred focus of Easton's piratical activities. His maraudings in Newfoundland

---

[12]Todd Gray, "Turkish Piracy and Early Stuart Devon," *Reports and Transactions of the Devonshire Association for the Advancement of Science, Literature and Art*, CXXI (1989), 159-171.

[13]Citing Prowse, *History*, Clive Senior states that a squadron of Turkish pirates in 1625 captured twenty-seven ships and took two hundred prisoners in Newfoundland; Senior, "Confederation," 345. There is nothing in Prowse, however, to suggest that this attack took place *in* Newfoundland, only that it involved ships *from* Virginia and Newfoundland. Todd Gray suggests that at least one Dartmouth ship was taken in Newfoundland; Gray, "Turkish Piracy," 164. This, however, was an exception to the general rule.

[14]Senior, "Confederation," 352n; E. Hunt, "Peter Easton," in *DCB*, I, 300-301; Gillian Cell (ed.), *Newfoundland Discovered: English Attempts at Colonization, 1610-1630* (London, 1982), 8-9 and 81-82; and Quinn (ed.), *New American World*, IV, 150. Other pirates included Captain Saxbridge and Henry Mainwaring; Senior, "Confederation," 340-341.

[15]Senior, "Confederation," 340. According to Senior, Easton "forced about 500 British fishermen to join him" while Mainwaring "pressed about one sixth of the British fishing fleet into his service." *Ibid.*, 340-341.

[16]Richard Whitbourne, "A Discourse and Discovery of New-Found-Land (1622)," reprinted in Cell, *Newfoundland Discovered*, 101-206, esp. 113.

were therefore intended to build up the strength of his force before sailing for the Azores where he intended to intercept the Spanish treasure fleet.[17] In short, "Newfoundland...offered a possible alternative to Ireland as a 'Nursery and Storehouse' for pirates."[18]

Despite this appeal, piracy at Newfoundland gradually faded as a serious problem after the middle of the seventeenth century. Several developments, all relating to the increased presence of instruments of the state for protecting and regulating the fisheries, may have contributed to this decline.[19] One such development was the state navy; in 1649, for the first time, the British government assigned warships to escort convoys of fishing fleets out to Newfoundland, then back to England or to their market destinations; before long, convoys became a regular feature of the trade.[20] Once in Newfoundland, the escort vessels remained on station for the duration of the fishing season. This system of "stationed ships" developed into the cornerstone of England's strategy for the protection of the Newfoundland fishery.[21] The authorities in London did not perceive Newfoundland as an overseas possession like other colonies, but rather, as one official would later explain, as "a great English Ship moored near the Banks during the Fishing Season, for the convenience of the English Fishermen."[22] In a migratory fishery there would be nothing left on

---

[17]John Guy and Richard Holworthy in Quinn (ed.), *New American World*, IV, 150-151; and Whitbourne, "Discourse," in Cell (ed.), *Newfoundland Discovered*, 113.

[18]Senior, "Confederation," 340.

[19]Yet piracy flourished in the Caribbean after mid-century. The rapid expansion there of the sugar-and-slave economy at this time created conditions which, in conjunction with the other factors mentioned here, may have caused pirates to lose interest in Newfoundland.

[20]Gillian Cell, *English Enterprise in Newfoundland, 1577-1660* (Toronto, 1969), 120-121. By the 1670s and 1680s, the Royal Navy was deploying both convoy escorts and cruising squadrons in the Mediterranean and its approaches for the protection of British commerce, including the Newfoundland fish trade, against the Barbary corsairs. See Sari Hornstein, *The Restoration Navy and English Foreign Trade, 1674-1688* (Aldershot, 1991); and Patrick Crowhurst, *The Defence of British Trade 1689-1815* (Folkestone, 1977).

[21]Gerald S. Graham, "Newfoundland in British Strategy from Cabot to Napoleon," in R.A. MacKay (ed.), *Newfoundland: Economic, Diplomatic and Strategic Studies* (Toronto, 1946), 245-264, esp. 258-259.

[22]Great Britain, Privy Council, *In the Matter of the Boundary between the Dominion of Canada and the Colony of Newfoundland in the Labrador Peninsula* (London, 1926-1927), Vol. IV, Joint Appendix, Part X, No. 722, "Extracts from Evidence

the island to protect once the fishing season was over; the capital investment, the labour force, the entire production of the fishery all abandoned the island in the fall and returned to Europe. The expense of fixed fortifications and permanent garrisons was not justified in such a place. The Committee of the Privy Council accordingly concluded in 1675 that fortifications were unnecessary in Newfoundland, declaring that "tis needless to have any such defence against Foreigners, the Coast being defended in the Winter by the Ice, and [is defended] in Summer by the resort of your Majesties Subjects, for that place will alwayes belong to him that is superior at Sea."[23] The "permanent" population at the time was less than two thousand and so widely dispersed that, in the opinion of the Lords of Trade, "no fortifications can be any security, by reasons of the distance of the Harbours."[24]

The opinions of the Privy Council and the Lords of Trade notwithstanding, a number of the larger communities did undertake to provide themselves with fortifications.[25] Moreover, France did not share England's objections to settlement or fortifications in the fishery, having established the fishing colony of Plaisance, or Placentia as it was known to the English, in 1662. Over the years, Plaisance acquired a fairly elaborate system of harbour fortifications supported by a military garrison.[26] Late in 1696, these were used as a springboard by Pierre Lemoyne d'Iberville to launch a devastating series of

---

of William Knox, given before Committee Appointed to Inquire into the State of Trade to Newfoundland, 1793."

[23]W.L. Grant and James Munro (eds.), *Acts of the Privy Council of England: Colonial Series* (6 vols., London, 1908-1912), I, 622, "Report of the Committee of the Privy Council, 15 April, 1675." Unless specifically indicated to the contrary, the dates on English documents are given in the Julian or Old Style, except that the year is taken to begin on 1 January, not 25 March as was then customary. This puts the dates of English documents eleven days behind those of French documents which are given in the Gregorian style.

[24]Gerald S. Graham, "Britain's Defence of Newfoundland; A Survey, from the Discovery to the Present Day," *Canadian Historical Review*, XXIII, No. 3 (1942), 260-261.

[25]Olaf Janzen, "New Light on the Origins of Fort William at St. John's, Newfoundland, 1693-1696," *Newfoundland Quarterly*, LXXXIII, No. 2 (1987), 24-31.

[26]Jean-Pierre Proulx, *Placentia, Newfoundland: The Military History of Placentia: A Study of the French Fortifications (and) Placentia, 1713-1811* (Ottawa, 1979); Frederick John Thorpe, "Fish, Forts and Finance: The Politics of French Construction at Placentia, 1699-1710," Canadian Historical Association, *Historical Papers 1971* (Ottawa, 1972), 52-63.

overland raids on the English fishing settlements.[27] Consequently, and despite their previous reservations, the English government also authorized the construction of fortifications, complete with a permanent garrison, at St. John's.[28] Neither the French defences at Plaisance nor the English defences at St. John's had the desired effect of discouraging attack on their respective fisheries.[29] The efforts to fortify the centres of the fishery in Newfoundland were, however, indications of the way in which the size, power and authority of the European state were not only growing in the second half of the seventeenth century but also were being extended overseas. And it was the expansion of the power of the state that contributed to the retreat of piracy during the second half of the seventeenth century.[30]

While the growth in size and service of the Royal Navy and the increased willingness to invest in overseas fortifications were two of the more visible manifestations of the rise of the modern state, the systems of trade and colonial regulation, law and jurisprudence and international diplomacy which began to make their appearance and which the navy enforced were probably of greater significance in the eventual suppression of piracy.[31] Some, like the Navigation Acts which emerged at mid-century, bureaucratic developments like the evolution of the Board of Trade by 1696 or the persistent efforts to extend the Vice-Admiralty court system to overseas colonies, became the foundation of what would later be identified as mercantilism.[32] Others, like the legislation known as the statute 10 & 11 William III c. 25 and more commonly

---

[27]Alan F. Williams, *Father Baudoin's War: D'Iberville's Campaigns in Acadia and Newfoundland 1696, 1697* (St. John's, 1987).

[28]James Candow, "The British Army in Newfoundland, 1697-1824," *Newfoundland Quarterly*, LXXIX, No. 4 (1984), 21-28.

[29]Fort William, as the principal work at St. John's was named, did not discourage French attacks on that town in 1705 and 1709; *ibid.*, 22.

[30]Ritchie, *War against the Pirates*, 143-157.

[31]Walter Minchinton provides a succinct discussion of the relationship between "an increased regard for the international rule of law" and the decline of piracy in "Piracy and Privateering in the Atlantic, 1713-76," another paper presented at the 1975 ICMH colloquium in San Francisco on "Piracy and Privateering."

[32]Ritchie, *War against the Pirates*, 148-153. After an unsuccessful first attempt in the 1670s and early 1680s, Vice-Admiralty courts were authorized by the statute 11 & 12 William III c. 7 (1698-1699) in the North American colonies and the West Indies to try cases of piracy, rather than have the accused sent back to England for trial; Ritchie, *War against the Pirates*, 144-146 and 150; and Minchinton, "Piracy and Privateering."

known as the Newfoundland Act of 1699, applied specifically to Newfoundland. All played their part in the decline of piracy in the North Atlantic late in the seventeenth century. Piracy persisted, but it was increasingly forced to retreat to more peripheral areas like the African coast or the Indian Ocean where the authority of the European state could not be exercised quite so vigorously.[33]

Nevertheless, and despite this trend, piracy experienced a sharp and dramatic revival after 1713, the year which brought the War of the Spanish Succession to a close. According to Marcus Rediker, "Anglo-American pirates created an imperial crisis with their relentless and successful attacks upon merchants' property and international commerce between 1716 and 1726." He estimates that between 1716 and 1726, "some 4,500 to 5,500 men went...'upon the account.'"[34] Traditionally, this resurgence of piracy was attributed to the effects of the War of the Spanish Succession. Unemployment was widespread as navies demobilized and as privateering came to an end.[35] The ensuing labour surplus in the maritime trades caused wages to drop and working conditions to deteriorate.[36] As Captain Johnson warned, "such Usage breeds Discontents amongst them, and makes them eager for any Change."[37]

---

[33]Ritchie, *War against the Pirates*, 32. According to Rediker, pirates located themselves "as distant as possible from the powers of the state." Marcus Rediker, *Between the Devil and the Deep Blue Sea: Merchant Seamen, Pirates, and the Anglo-American World, 1700-1750* (Cambridge, 1987), 257.

[34]Rediker, *Devil*, 254-256. See also Ritchie, *War against the Pirates*, 233ff. According to Stanley Richards, "New converts to piracy declared themselves as going *On the Account*. It merely meant that they would not be drawing wages, as in...a merchantman, but a nominal share...when plunder was taken." Stanley Richards, *Black Bart* (Llandybie, 1966), 36. Richards' biography is based largely on Captain Charles Johnson, *A General History of the Pyrates: From Their First Rise and Settlement in the Island of Providence, to Present Time* (London, 1724; reprint, Charleston, SC, 2012), 4. The editor of the original volumes, Manuel Schonhorn, who edited the reprint, accepts the view that this work was penned by Daniel Defoe under the pseudonym of Captain Charles Johnson. Today, however, Johnson is generally recognized as a separate individual.

[35]The Royal Navy was reduced in size from 49,860 men in 1713 to 13,475 in 1715 and 6298 in 1725; Rediker, *Between the Devil and the Deep Blue Sea*, 281. The perception that unemployed privateersmen turned frequently to piracy occasioned Captain Johnson's condemnation of "Privateers in Time of War [as] a Nursery for Pyrates against a Peace." Johnson, *General History*, 4.

[36]Rediker, *Between the Devil and the Deep Blue Sea*, 282. According to Rediker, wages in 1713 were half those in 1707.

[37]Johnson, *General History*, 4.

Piracy, concluded Rediker, was not motivated so much by a desire to get rich quick but by the "jarring social and economic effects" that followed the end of the war in 1713 and particularly by a desire "to revenge themselves on base Merchants and cruel commanders of Ships."[38]

Pirates would have found the Newfoundland fisheries at this time to be a particularly fertile recruiting ground, for working and living conditions there were extremely difficult. The wars of 1689 to 1713 had disrupted not only the markets for Newfoundland fish in Europe but also the transatlantic movement of men and ships which was the definitive characteristic of the migratory ship fishery. A population of permanent inhabitants or "planters" became an important element of the fishery, as more and more ships came from England not simply to fish but to trade, depending on the planters to complete their cargoes. Under wartime conditions the price of fish rose, but so did the cost of provisions and supplies needed to survive the long Newfoundland winter.[39] Credit-dependency and debt became increasingly characteristic, both of the planter's relationship with his supplier and of the planter's relationship with his employees or "servants." Servants were paid in kind (or "truck"), when they were paid at all.[40] Many found themselves stranded in Newfoundland, unable to return home, yet with no resources by which to support themselves. Adding to their woes after 1711 was a mysterious alteration in the migratory pattern of the cod: the fish refused to come inshore, causing catches to decline from a wartime high of 400 quintals per boat to below 200 quintals.[41] The fishery in 1712 experienced its most unproductive season ever. Countless merchants went bankrupt and many planters were left "irretrievably in debt."[42] As word spread back to England, men refused to hire themselves out into the fishery. A labour shortage ensued within the fishery, forcing wages up, but the high wages were a chimera; planters who offered them did so knowing that there was little likelihood that the wages would be paid. At best, the trend to-

---

[38]William Snelgrove, cited in Rediker, *Between the Devil and the Deep Blue Sea*, 271-272 and 281-282. Snelgrove was seized by pirates at the mouth of the Sierra Leone River in 1719 and spent a month in their company.

[39]Matthews, "History," 244-247, 248-249, 256-258 and 276-278.

[40]*Ibid.*, 276-279 and 287ff. Peter Pope discusses the late seventeenth-century appearance of the truck system in "Historical Archaeology and the Demand for Alcohol in 17th Century Newfoundland," *Acadiensis*, XIX, No. 1 (1989), 87-88.

[41]Matthews, "History," 306-307.

[42]*Ibid.*, 301-304 and 314-318.

wards debt and payment in kind accelerated.[43] Many servants stranded in New-foundland chose to escape by selling themselves into labour contracts and emi-grating to New England.[44] Those who remained were receptive to the appeal of piracy.

Pirates, of course, normally gravitated to the busy – and therefore lucrative – trade lanes of the Caribbean and South America. The value and ease of disposing of the cargoes to be found there, the myriad number of is-lands and the multinational character and lack of effective local authority over many of the islands all gave the West Indies their appeal to pirates. The trade of the Anglo-American coast, Africa and the Indian Ocean also attracted pi-rates. Nevertheless, Newfoundland was not immune to the revival of piracy, notwithstanding the more prosaic nature of its trade and its fishing economy.[45] Almost as soon as piracy revived elsewhere, the merchants trading to New-foundland were petitioning government for protection, alleging that pirates were also very numerous in the waters off southern Newfoundland.[46] But it was not until Bartholomew Roberts made his brief appearance in 1720 that piracy became a serious matter in Newfoundland.

The year before, in 1719, Bartholomew Roberts (1682?-1722) had been third mate of the *Princess* galley, Captain Abraham Plumb, when the ship was captured by pirates led by Howel Davis as it picked up a cargo of slaves at

---

[43]According to Keith Matthews, a boat-master's seasonal wage, which was £12-14 in 1709, had risen to £10-30 in 1715; by 1717 even "common" fishermen were being offered £20. *Ibid.*, 315-317.

[44]According to Captain William Passenger, who commanded the Newfound-land station in 1717, as many as 1300 men were believed to have been carried off to New England. See Canada, Library and Archives Canada (LAC), Colonial Office (CO) 194, vol. 6, 199-200, microfilm B-209, Passenger to Board of Trade, 5 October 1717.

[45]The cargoes of most fishing vessels were not worth much. A French bank-ing vessel captured by William Kidd in 1696 was subsequently condemned in New York as a prize; the ship and its cargo of fishing tackle and salt fetched £312. See Ritchie, *War against the Pirates*, 62 and 257n. Contrast that with the cargo worth £8000 which fell into Bartholomew Roberts' hands when he captured *Samuel*, Captain Samuel Carry, as the pirates were making their way south following their descent on Newfoundland; *CSPC*, XXII, No. 200, Governor Shute (Boston) to Board of Trade, 19 August 1720. A full account of the capture of *Samuel* appears in Johnson, *General History*, 217.

[46]Great Britain, National Archives (TNA/PRO), Admiralty (ADM) 1, vol. 4099, 387-390, Secretary Methuen to the Lords of the Admiralty, 19 March 1716/1717, with enclosures from the Lords of Trade to Secretary Methuen, 27 Febru-ary 1716/1717 and a petition sent by the merchants of Bideford requesting that a war-ship be ordered to cruise between Trepassey and St. Peter's (the island of St. Pierre) to protect the trade against pirates.

Annamabo on Africa's Gold Coast.[47] Roberts and a number of his shipmates were conscripted by the pirates as was then often the case, for the high rate of mortality made the acquisition of new recruits, both willing and unwilling, a matter of constant urgency. Within a matter of weeks, the erstwhile conscript found himself elected captain and leader by his captors, Captain Davis having been killed.[48] Roberts appears to have had a talent for his new profession, for under his leadership the pirates captured several vessels in Africa, then several more, including some rich prizes, on the coast of South America. He lost much of his loot when he and part of his crew became separated from his ship, which decamped without him. Undaunted, Roberts made for the Windward Islands where he seized several small sloops and began rebuilding his crew. When local French and English authorities organized counter-measures, Roberts headed north, eventually ending up in Newfoundland early in June 1720.[49]

Roberts' biographer seems surprised that the pirates would reveal their presence in Newfoundland by first attacking shipping on the banks before making for a harbour where they could rest and refit.[50] Yet Roberts' strategy was an astute one. With only a ten-gun sloop and a crew of about forty-five men, Roberts lacked the strength to face down a serious challenge from his intended victims.[51] By capturing several of the small vessels that prevailed in

---

[47]Richards, *Black Bart*, 19-23. See also Johnson, *General History*; and J. Franklin Jameson (ed.), *Privateering and Piracy in the Colonial Period: Illustrative Documents* (New York, 1923; reprint, New York, 1970).

[48]It was quite common for pirates to elect their leaders, often on the basis of intangibles such as luck and skill in capturing prizes. A captain exercised authority in combat but enjoyed few privileges, for a pirate ship functioned according to a rude democracy, the men voting on critical decisions, binding themselves to a set of articles or "compacts" which defined the rules and customs of the ship, and frequently insisting that power be divided between the captain and an elected quartermaster who represented and protected the interests of the crew; Rediker, *Between the Devil and the Deep Blue Sea*, 261-263. This certainly appears to have been the case with Bartholomew Roberts, whose pirate crew "permitted him to hold this office only on condition that they as a body might be master over him." Richards, *Black Bart*, 37. The articles of the compact also defined the distribution of plunder by shares; according to Rediker, this reveals that "pirates did not consider themselves wage labourers but rather risk-sharing partners." Rediker, *Between the Devil and the Deep Blue Sea*, 264.

[49]Richards suggests that Roberts sailed north to sell in the North American colonies the various goods he had captured; Richards, *Black Bart*, 41. If so, there is no evidence to confirm this.

[50]*Ibid.*, 41.

[51]In a despatch to the Board of Trade dated 31 May 1721, Lieutenant Governor Spotswoode of Virginia claimed Roberts arrived at Trepassey with a crew of sixty

the French and English bank fisheries, Roberts generated a fear that became his most effective weapon in the next step of his foray – the occupation of Trepassey on the southeastern tip of Newfoundland.[52] In late June or early July, the pirate sloop entered the harbour there "with Drums beating, Trumpets sounding, and other Instruments of Musick, English Colours flying, their Pirate Flagg at the Topmast-Head, with Deaths Head and Cutlash."[53] The sight and the noise triggered a panic as crews abandoned nearly two dozen fishing vessels moored there in their efforts to flee the pirates.

It is impossible to determine whether Roberts' choice of Trepassey was deliberate or fortuitous. It was certainly a sensible one in 1720 for a number of reasons. Its proximity to the great fishing banks and to the trade lanes between the New World and the Old gave it strategic value to the pirates. Furthermore, while the English had been slow to develop the former French settlements on Newfoundland's south coast, Trepassey was an exception. English fishing vessels, which had abandoned Trepassey during the recent war, had returned there by 1715, and by 1720 had developed it into a major centre of the English migratory bye-boat and bank fisheries.[54] Roberts could therefore find everything he needed at Trepassey to refit his ship. He could also do so in reasonable security, for the harbour had no defences of its own and the warships stationed in the fishery rarely appeared at Trepassey.[55] In 1720, there

men; *CSPC*, XXXII, No. 513. According to the *Boston News-Letter*, 22 August 1720, Roberts had only forty-five men; see the extract in Jameson (ed.), *Privateering and Piracy*, No. 117, 317.

[52]According to Moses Renos (or Renolds), a mariner of Dartmouth, Roberts seized the pink on which he was serving, together with four or five more vessels, before making for Trepassey. Renos was forced to join Roberts' crew until released at the island of St. Christopher's in September. See *CSPC*, XXXII, deposition of Moses Renos, enclosure No. 251 (iv) in Governor Hamilton (Antigua) to the Board of Trade, 3 October 1720, in *CSPC*, XXXII, No. 251. In his biography of Roberts, Stanley Richards emphasized that the "whole strategy [of pirates] was shaped to obtain wealth without having to fight for it. Thus they sought to frighten their victims into surrender;" Richards, *Black Bart*, 22.

[53]Extract from the *Boston News-Letter*, 22 August 1720 in Jameson (ed.), *Privateering and Piracy*, No. 117, 317.

[54]Tom Nemec, "Trepassey, 1505-1840 A.D.: The Emergence of an Anglo-Irish Newfoundland Outport," *Newfoundland Quarterly*, LXIX, No. 4 (1973), 17-28.

[55]Governor Spotswoode later claimed that there were "upwards of 1200 men and 40 p[iece]s of cannon" when Roberts appeared, yet Roberts could act with impunity "for want of courage in this headless multitude." CSPC, XXXII, No. 513, Spotswoode to Board of Trade, 31 May 1721. This remark is typical of the exaggerated claims that were subsequently made to emphasize the seriousness of Roberts' raid; the summer

were only two warships – the aging fourth-rate *St. Albans* (50), Captain Francis Percy, and a sixth rate, almost certainly a twenty-gun frigate, the smallest in its class.[56] It was then normal practice for both ships to spend most of their time in port, *St. Albans* at St. John's and the frigate probably at Placentia, while the essential task of supervising the inshore fishery was left to small boats.[57] Consequently, during the two weeks or so that Roberts was at Trepassey, no warship ever made an appearance. Nor was the nearest garrison, also at Placentia, in any position to respond. Not only did it lack overland and sea communication with Trepassey, it was in the course of being reduced in size in 1720. Indeed, the garrison commander, Colonel Samuel Gledhill, used the pirate raid to argue for an expansion of his command, arguing that with a larger garrison he could construct a military trail to connect Placentia with adjacent outports like Trepassey.[58]

Roberts was therefore free to consolidate his control over Trepassey with impunity. He does not appear to have caused very much destruction, burning only one vessel and generally leaving the rest alone.[59] Instead, like

---

population was half that claimed by Spotswoode, while there is no evidence that there were any cannon whatsoever at Trepassey.

[56]*St. Albans* was built at Deptford in 1688; William Sutherland, "Reflections on Shipbuilding, 1660-1714," in John B. Hattendorf, *et al.* (eds.), *British Naval Documents, 1204-1960* (Aldershot, 1993), No. 160, 268.

[57]The senior commanding officer of the two stationed ships was designated Commander-in-Chief of the Newfoundland station, and to him would also be assigned the civil powers of governor. It was these responsibilities which normally bound him to St. John's, located at the centre of the British fishery at Newfoundland.

[58]LAC, CO 194/6, 367, Samuel Gledhill to Board of Trade, 3 July 1720.

[59]Captain Johnson insisted that "It is impossible...to recount the Destruction and Havock they made here, burning and sinking all Shipping, except a *Bristol* galley, and destroying the Fisheries, and stages of the poor Planters, without Remorse or Compunction; for nothing is so deplorable as Power in mean and ignorant Hands." Johnson, *General History*, 216. It was an assessment repeated by Roberts' biographer; see Richards, *Black Bart*, 42. Johnson's account, in turn, echoes information in *CSPC*, XXXII, No. 200, Governor Shute to the Board of Trade, 19 August 1720 (Shute was convinced that the fishery at Newfoundland had been destroyed); *CSPC*, XXXII, No. 513, Governor Spotswoode (Virginia) to the Board of Trade, 31 May 1721; and the extract of 22 August 1720 in the *Boston News-Letter* reprinted in Jameson (ed.), *Privateering and Piracy*, 317. Samuel Gledhill, the Lieutenant Governor of Placentia, also claimed, without supporting evidence, that the pirates "have Burnt and destroyed 26 Ships with great Numbers of ffishing Craft." LAC, CO 194/6, 367, Gledhill to the Board of Trade, 3 July 1720. On the other hand, the only eyewitness account, the deposition of Moses Renos, denies that there was very much destruction; see *CSPC*, XXXII,

Peter Easton more than a century before, he used Newfoundland as a sort of "service centre," a vast chandler's shop, fully stocked with provisions, supplies and nautical stores, a veritable cornucopia of goods necessary to keep a pirate ship afloat.[60] He did seize one vessel to replace his little sloop, forcing the crews of the fishing ships to refit and arm the vessel with eighteen guns. Roberts also recruited a number of the men into his crew. Finally, the pirates requisitioned provisions, for it was still early in the season, and the provisions brought by the fishing ships from England would not yet have been depleted. Thus re-stocked with a larger ship, additional men, arms and provisions, Roberts headed out to the fishing banks for a brief cruise. Here he managed to capture a twenty-six-gun French vessel, which he promptly took in place of the vessel so labouriously re-fitted in Trepassey. Several other French and English vessels were also seized and pillaged before Roberts finally set his course south to return once again to the Caribbean.[61] Notwithstanding the reports of various colonial administrators, Trepassey was left behind, none the worse for wear, having served its brief but crucial purpose of refitting and re-supplying the pirates.[62]

Piracy persisted in Newfoundland for a few more years. In 1721, the bank fishery was barely underway when it was again disrupted, this time by a trio of French pirate ships (of eighteen, twelve and twelve guns, respectively) reportedly working together under the leadership of a pirate named Laubé. They seized several French bank ships, stripping them not only of clothing,

---

No. 251(iv), enclosure with Governor Hamilton (Antigua) to the Board of Trade, 3 October 1720. And in his report later that year to the Board of Trade on the state of the fishery, the commodore of the Newfoundland station, Captain Percy, said not a word about Bartholomew Roberts, the raid on Trepassey or pirates, suggesting that the raid by Roberts was far from devastating; LAC, CO 194/7, 5-10n, Capt. Percy, "Answers to the Heads of Inquiry," 8 October 1720. Nor were the "Returns of the Fishery" at Trepassey significantly worse in 1720 than in 1722; LAC, CO 194/7, 12.

[60]According to Matthews, "History," 368, the ships of Poole by 1714 were carrying to Newfoundland "cotton goods, hats, stockings, oats, bread, beef, pork, nails, wheat, cordage, peas, bacon, leatherware, hardware, cabbage and woollen goods."

[61]It was as he began making his way south from the fishing banks that he stumbled across *Samuel*. See above, note 45.

[62]Trepassey was not entirely unaffected by Roberts' visit. When a French barque was forced into Trepassey by bad weather in 1722, it was cannonaded by the English, its captain and crew imprisoned, its fishermen detained and its goods seized. Archives of Fortress of Louisbourg National Historic Park, transcripts of documents in Paris, Archives Nationales, Colonies (hereafter AFL AN Colonies), série C^{11}B/6, 134-134v, M. Dauteuil to the Minister of Marine, 29 October 1722.

arms, powder and provisions but also of their fishing gear, cordage and tackle. Invariably, the pirates ignored the small quantities of fish that had been caught to that point. Instead, every vessel lost a substantial part of its crew, always young sailors; one lost more than half of its crew of twenty, another lost every man but two. In at least three instances, they also took the ship's chirurgeon or surgeon together with his equipment and the carpenter's equipment as well – skills and equipment which were of great value to men whose pirate status denied them the opportunity to put into port to attend to injuries or repairs. After cutting the shrouds and damaging their masts, the pirates ordered the fishing vessels to return to France; in some instances, vessels were sunk. The alarm quickly spread, and some bank ships returned to France rather than risk losing their investment or their men.[63]

It is impossible to determine how willingly the fishermen went who were taken by the pirates. According to witnesses, many who joined Bartholomew Roberts at Trepassey did so voluntarily. Presumably the harsh conditions within the fishery after 1713 made recruiting fairly easy. In at least one instance, recruits did not wait to be asked. In August 1723, one John Phillips, employed at St. Peter's as a splitter in a shore crew, conspired with some others to steal a schooner and become pirates. They quickly captured several small French and English banking vessels from which they added to their numbers. They then made their way south to the West Indies where they cruised through the winter.[64] In the spring, Phillips began making his way back to Newfoundland, possibly because his vessel was in a bad state of repair.[65] He plundered and robbed several New England and Newfoundland vessels on his way. This proved to be his downfall; in mid-April, some of his prisoners rose

---

[63]See Archives Nationales, Marine (AN Marine), série B³/269, 185-191, 207-209v and 230-231v, M. de Silly (*Ordonnateur*, Le Havre) to Minister of Marine Maurepas, 5 and 27 September and 29 October 1721; see also AN Marine, B³/269, 524-525, de Champmort (*capitaine de vaisseau* and *Commandant de la Marine* at Port Louis) to Maurepas, 10 November 1721.

[64]The story of John Phillips is recounted in Johnson, *General History*, 341-350; and, through documents numbered 119 to 124, consisting of affidavits and transcripts, in Jameson (ed.), *Privateering and Piracy*.

[65]The pirate vessel was described by Joseph Galpin, master of *St. Charles* of Nantes, as a schooner of thirty or forty tons, seventeen or eighteen men, armed with two small cannon and eight smaller pieces of ordnance. The mizzen-mast had been fished and in a very bad state ("*en tres mauvais Estat*"). Galpin had been making his way back to France from the West Indies when he ran into Phillips, who by then was returning to Newfoundland. Phillips pillaged *St. Charles*, roughed Galpin up and took four of his crew. See AN Marine, B³/295, 165-167v, extract of a declaration made by Galpin on 4 May 1724 before the Bureau des Classes at Nantes, in Bigot de la Mothe (*commissaire-générale*, Nantes) to Minister of Marine Maurepas, 9 May 1724.

up and overpowered the pirates; Phillips was killed and the pirate schooner was carried into Boston.[66]

By then, the resurgence of piracy in Newfoundland was reaching its climax. In 1724, a fairly substantial pirate ship, in company with two or three vessels of lesser force, terrorized the banking fleets as well as the inshore fisheries on Newfoundland's south coast.[67] The fishing interests in England, France and Newfoundland protested that the fishery was threatened with complete destruction, owing largely to the lack of adequate protection, particularly in the early part of the year before the station ships had arrived.[68] Yet steps *were* being taken to deal with the problem. Beginning in 1722, London assigned three warships (including a second fourth-rate) instead of the customary two to the Newfoundland station. By 1724, at least one of the ships patrolled the banks regularly, and sometimes there were two.[69] France, which had never previously assigned warships to patrol the fisheries, now ordered at least two and possibly more to patrol the banks.[70] The French warships promptly cap-

---

[66]Johnson, *General History*, 346-350.

[67]A pirate of thirty-six guns and 200 men "*de toutes nations mais principallement anglois*" captured and pillaged some French bankers in May 1724; AN Marine, B³/295, 192-193, de la Mothe to Maurepas, 17 June 1724.

[68]"*Les forbans...ruinent la pêche.*" AFL AN Colonies, C¹¹B/6, 199-204v, Governor Saint-Ovide (Louisbourg) to Minister of Marine, 26 November 1723. The planters and fishing shipmasters at Placentia complained that "the Pirates being Sencible that there has not been any man of warr, Stationed for this Place; they every Year, come Early in the Spring...and frequently disturbs us, and makes us fly from...our Fishery." TNA/PRO, ADM 1/2453(i), Capt. St. Lo to Secretary to the Admiralty Burchett, 14 November 1724.

[69]*Argyle* (50), Captain Robert Bouler was on the banks in June; TNA/PRO 1/1473, Bouler to the Admiralty, 13 August 1724. When Bouler learned that *Ludlow Castle*, Captain St. Lo, had set out from Placentia in late July in search of a fourteen-gun pirate vessel, he ordered *Solebay*, Captain Knighton, to cruise as well; TNA/PRO, ADM 1/1473, St. Lo to Knighton, 12 September 1724.

[70]There may have been as many as five French warships on the banks in 1724. *Herculle* and *Prothée*, under the command of M. de Rocquefeuille, were ordered to the banks that year to cruise against pirates; they left Brest in April. AN Colonies, B/47, 1268, Maurepas to Saint-Ovide and Le Normant de Mézy, 6 June 1724; and AN Marine B³/295, 209, de la Mothe to Maurepas, 1 July 1724. In June, *Argyle*, Capt. Bouler, spoke to two French warships, possibly de Rocquefeuille's, on the banks; he described them as "One of 60, and the Other of 54 Guns." TNA/PRO, ADM/1473, Bouler to the Admiralty, 13 August 1724. Bouler learned that another French warship of forty guns was cruising off Cape Breton Island. Finally, a French banker encountered the frigates *Argonaute* and *Amazonne*, "*armé a Brest*" as well as "*une fregatte du*

tured the largest of the pirate ships and sank another.[71] When a particularly vicious English pirate attacked the bank fleet again in 1725, both French and English warships responded quickly.[72]

Such measures and actions may have had the desired effect; by 1726, piracy appears to have disappeared as a serious problem in the fishery. Yet piracy everywhere in the North Atlantic was in retreat by then, and if the vigour of naval campaigns against the pirates deserves some of the credit for this trend, then so does the well-publicized execution of pirates like those in John Phillips' crew after they were taken into Boston for trial, and the introduction of harsh laws to criminalize all collaboration and contact with pirates.[73] With piracy fading elsewhere, Newfoundland's role as a "service centre" faded as well. Pirates had been drawn to Newfoundland primarily for two reasons which no longer applied after 1725. The protective hand of government, once practically non-existent in the fishery, had become both more visible and more energetic. If pirates gravitated to places where defences against them were weak and the navy non-existent, then Newfoundland held little appeal for them after the mid-1720s.[74] And though the harsh conditions within the fishery had enhanced Newfoundland's appeal as a recruiting ground, particularly during the decade after 1713, the gradual recovery of the fishery after the mid-1720s and the introduction after 1729 of a rudimentary system of civil administration

---

port de Rochefort" cruising the banks in search of pirates; AN Marine B³/293, 305ff, declaration of Capt. Jean Montard of Granville, master of the banker *Phylypeaux*, in de Silly and de Villiers to Maurepas, 28 August 1724. These frigates may have been destined elsewhere and were simply adding their strength to *Herculle* and *Prothée* while crossing the banks.

[71]AN Marine B³/293, 310-311, declaration of Capt. Guillaume, Justice of Granville, master of the ship *Marie de Grace*, 31 August 1724, in de Silly and de Villiers to Maurepas, 31 August 1724.

[72]*Elisabeth* and *Jazon* were ordered that summer to patrol the banks; AN Marine B³/294, 150-151, M. Morin (*commissaire de Marine*, Saint-Malo) to Maurepas, 31 December 1724. *Ludlow Castle*, Captain John St. Lo, was stationed at Placentia and spent much of the summer cruising in search of pirates. While pirates frequently roughed up their victims, the English pirate reported in 1725 had no compunction about shooting them dead; AN Marine B³/301, 408-409, du Hallier (*commissaire des classes*, Granville) to Maurepas, 20 August 1725. The pirate may have been "Spriggs." Captain St. Lo attempted to disguise *Ludlow Castle* "to Decoy Spriggs a Pirate who is report^d to be On the Banks." TNA/PRO, ADM 51/588(iv), captain's log, *Ludlow Castle*, Capt. John St. Lo.

[73]Rediker, *Between the Devil and the Deep Blue Sea*, 282-283.

[74]Ritchie, *War against the Pirates*, 32.

and jurisprudence would have diminished that role as well.[75] Newfoundland's brief experience as an eighteenth-century "nursery and storehouse for pirates" had ended.

---

[75]Christopher English, "The Development of the Newfoundland Legal System to 1815," *Acadiensis*, XX, No. 1 (1990), 89-119.

A Scottish Sack Ship in the
Newfoundland Trade, 1726-1727[1]

Early in the morning of 30 July 1726, the ship *Christian*, Captain Alexander Hutton, passed through the Narrows into the harbour of St. John's, Newfoundland.[2] In its hold were thirty-two casks holding five tons of "Bisquett."[3] As well, the vessel carried a substantial quantity of fabrics, clothing, shoes and a variety of sundry goods ranging from inkhorns to buckles and buttons to thimbles and pins.[4] The investors who had chartered *Christian* had directed the ship to "St Johns, Ferryland, or the Bay of Bulls, or any Harbour thereabouts," where the master and supercargo were instructed to sell the biscuit "to the best advantage, rather to Masters of Ships, than to the fishermen upon the Island." They were then to acquire "where you best can" a cargo of "good Merchantable fish, well dryed & fair to the eye," without "Spots or blemishes," for subsequent delivery to Barcelona, Spain. There the investors' agent would arrange for the sale of the fish and the acquisition of a partial cargo of cork before *Christian* began the homeward journey. The ship would stop only at Sanlúcar, north of Cádiz, to complete its cargo with sherry and fruit.[5] The voyage was not without its share of uncertainties. It might not find fish in Newfoundland to complete its cargo; it might find itself caught by the outbreak of the war that was then threatening between England and Spain; it might find

---

[1]This essay appeared originally in *Scottish Economic and Social History*, XVIII, Part 1 (1998), 1-18.

[2]Much of this paper is based on entries in the "Journal of a Voyage to Newfoundland, in 1726-1727 in *Christian* of Leith, Captain Alexander Hutton and Edward Burd Jr., Supercargo." The journal was kept by Burd and is in the Scottish Record Office (SRO), RH9/14/102. The dates provided in the journal are all in the Julian or Old Style, and unless indicated to the contrary, all references to dates in this paper are also in the Old Style, except that the year is taken to begin on 1 January, not 25 March as was then customary.

[3]10,150 lbs. of biscuit, or hard bread, in total; see Burd, "Journal," invoice of...Bisquett, 1-1v.

[4]These goods were carried on a number of speculative private accounts belonging to the ship's officers.

[5]Burd, "Journal," 10v-11, copy of instructions for Alexander Hutton, master of *Christian*, and Edward Burd Jr., supercargo, 24 May 1726.

itself a victim of piracy; it might not be able to pick up a cargo in southern Europe.[6] The investors' instructions were therefore quite specific on how to proceed in these eventualities. They knew that once at sea *Christian* was outside their control, and the success of their investment depended entirely upon the judgment and ability of the men they had placed in its charge.[7]

In its purpose, its cargo and its route, *Christian* was what was commonly called in the Newfoundland trade a "sack ship," one of at least fifty-five such ships reported in Newfoundland that year.[8] What set this particular voyage and its investors apart from the others was that *Christian* was from Leith, the port community serving the Scottish city of Edinburgh. Most sack ships were English, usually based in London or the West Country; recently, a few Irish ships had also ventured into the trade. A Scottish presence, however, was almost unprecedented. It is the purpose of this paper, in part, to determine what compelled the men who freighted *Christian* to venture into so unfamiliar a trade and in so doing to determine what significance, if any, can be attached to the voyage of this Scottish sack ship. That voyage was also unusual in the completeness of its business documentation, making this one of the best documented British sack ship voyages of the eighteenth century. This gives *Christian* an unusual capacity to reveal the full complexity and difficulties of this most important element in the Newfoundland fish trade.

---

[6]The resurgence of piracy in Newfoundland waters during the 1720s is examined in Olaf Uwe Janzen, "The Problem of Piracy in the Newfoundland Fishery in the Aftermath of the War of the Spanish Succession," in Poul Holm and Olaf Janzen (eds.), *Northern Seas Yearbook 1997* (Esbjerg, 1998), 57-75. Nevertheless, the greatest danger was posed by Algerine cruisers and "Barbary corsaires" in the Straits of Gibraltar and its approaches; Sari Hornstein, *The Restoration Navy and English Foreign Trade, 1674-1688* (Aldershot, 1991); and Patrick Crowhurst, *The Defence of British Trade, 1689-1815* (Folkestone, 1977).

[7]This effectively signified a power that the ship's master exercised over his employers. The trust that an employer bestowed upon the master might be given reluctantly, but it was essential given the limitations of early eighteenth-century transatlantic communications. See Ian K. Steele, "Instructing the Master of a Newfoundland Sack Ship, 1715," *Mariner's Mirror*, LXIII, No. 2 (1977), 191-193.

[8]Great Britain, National Archives (TNA/PRO), Colonial Office (CO) 194/8/42, copied in Library and Archives Canada (LAC), microfilm B-210, "State of the Fishery at Newfoundland," 1726. They were known as "sack ships" because they customarily engaged in the trade in sherry, or "sack," and other wines; Peter Pope, "Adventures in the Sack Trade: London Merchants in the Canada and Newfoundland Trades, 1627-1648," *The Northern Mariner/Le Marin du nord*, VI, No. 1 (1996), 1-19.

II

Sack ships first appeared late in the sixteenth century.[9] Until then, the so-called "fishing" ships that the English sent to Newfoundland usually brought their crews and the season's catch back to England.[10] The fish was then shipped to Iberia and southern Europe, which consumed almost all of the fish produced by the English fishery. This, however, was costly both in terms of the delay it entailed in getting the fish to market and in the additional handling it required. On the other hand, as Keith Matthews later observed, it was "hopelessly uneconomic" for the fishing ships to carry the season's catch directly to market because they were as small as their fishing crews were large.[11] In the year that *Christian* appeared in St. John's, ninety-four fishing ships totalling 5240 tons and 1943 men were reported at Newfoundland, for an average of just under fifty-six tons and nearly twenty-one men each.[12] The rapid expansion of the English fishery in the final quarter of the sixteenth century therefore coincided with the appearance of sack ships, which came to Newfoundland expressly to buy fish taken by others.

---

[9]Keith Matthews, "A History of the West of England-Newfoundland Fisheries (Unpublished DPhil thesis, Oxford University, 1968), 71 and 74. *Parnell* of London (1579/1580) appears to have been one of the earliest recorded English sack ships; Pauline Croft, "English Mariners Trading to Spain and Portugal, 1558-1625," *Mariner's Mirror*, LXIX, No. 3 (1983), 257.

[10]"So-called" because, in fact, the English fishing ships did not fish. Rather, they were moored as soon as they arrived in Newfoundland, and the men fished from small boats close to their fishing station. The British shift to the banks was pioneered by New England in the 1680s; Daniel Vickers, *Farmers and Fishermen: Two Centuries of Work in Essex County, Massachusetts, 1630-1850* (Chapel Hill, NC, 1994), 149-150. The British at Newfoundland followed this lead after 1713; Matthews, "History," 309-312; and C. Grant Head, *Eighteenth Century Newfoundland: A Geographer's Perspective* (Ottawa, 1976), 72-74.

[11]Matthews, "History," 71. Nevertheless, this did not preclude the fishing ship from participating in the trade. As Cell notes, "The larger sack ships...probably made the longer voyage more safely and more profitably, but even a fishing ship could reach the Levant as did the *Sunne* in 1600." Gillian Cell, *English Enterprise in Newfoundland, 1577-1660* (Toronto, 1969), 31.

[12]TNA/PRO, CO 194/8/42, "State of the Fishery at Newfoundland," 1726. This is considerably smaller than the seventy to 100 tons that was deemed the ideal size for a fishing ship at the beginning of the previous century; Cell, *English Enterprise*, 3. The development of the sack ship as a fixture in the Newfoundland fish trade undoubtedly contributed to the trend towards smaller fishing ships, since the fishing ship no longer needed to carry all its provisions out to Newfoundland or carry its complete catch to market.

These vessels were usually much larger than the fishing ships – the fifty-five sack ships at Newfoundland in 1726 totalled 5530 tons and carried 586 men, for an average of over a hundred tons each and less than eleven men per ship.[13] Nor did sack ships set out for Newfoundland until the fishing season was well underway (usually June or July), timing their arrival in the fishery for the moment when sufficient quantities of saltfish were at hand to provide a full cargo with a minimum of delay and immediate transportation to market. The aim was to be among the first to reach the markets, when prices were highest.[14] The smaller fishing ships continued in Newfoundland for another month or more, at which point they would either return to England with the fishing crews and carrying those Newfoundland products for which there was demand in England, such as fish oil, salmon and furs, or else they would transfer the bulk of their crews to other fishing ships, retaining sufficient men to handle the vessel, and carry whatever fish was on hand to market. In effect, the sack ship was a general freighter for which Newfoundland saltfish was simply one of several commodities carried in the course of a voyage about the North Atlantic and Mediterranean. For the fishing ship, in contrast, trade was merely an option; its principal role was to support a fishing expedition.

Sack ships acquired their cargoes of fish in one of two ways. Some received fish that had been purchased on consignment through arrangements made before the fishing season had begun.[15] This had commonly been the case early in the seventeenth century when the fishery was almost wholly a migratory one, in which the fishing ship was adequately provisioned, crewed and equipped for the entire season before it ever departed England. Without a significant number of resident fishermen to serve as a consumer market for whatever goods a sack ship might bring out, and to produce fish independently of the English fishing ships, the owners or charterers of the sack ship would want the security of a cargo of fish that had been arranged well in advance of the actual voyage. By the early eighteenth century, however, a small, resident

---

[13]TNA/PRO, CO 194/8/42, "State of the Fishery at Newfoundland," 1726. At seventy tons burthen, *Christian* was small for its role as a sack ship, but its size was fairly typical of Scottish ships of that era; Susan Mowat, *The Port of Leith: Its History and Its People* (Edinburgh, 1994), 288; and T.C. Smout, *Scottish Trade on the Eve of Union, 1660-1707* (Edinburgh, 1963), 47-52.

[14]Matthews, "History," 74; Cell, *English Enterprise*, 5-6 and 9-20; and John Gilchrist, "Exploration and Enterprise – the Newfoundland Fishery, c.1497-1677," in David Macmillan (ed.), *Canadian Business History: Selected Studies, 1497-1971* (Toronto, 1972), 7-26, esp. 19.

[15]See, for instance, the instructions given in 1634 to Thomas Breadcake, master of the sack ship *Faith* of London, in Ralph Davis, *The Rise of the English Shipping Industry in the Seventeenth and Eighteenth Centuries* (London, 1962; reprint, St. John's, 2012), 226-228; and Cell, *English Enterprise*, 18-19.

population had been planted in Newfoundland which was dependent upon a broad range of imported goods. The planters, along with a new category of migratory boat fishermen, known as "bye boatmen," and even the fishing ships themselves, relied increasingly on cargoes of provisions and dry goods brought to Newfoundland on a speculative basis by sack ships and trading vessels from British America. Under these conditions, more and more sack ships found it convenient to make their arrangements to purchase fish only after their arrival in Newfoundland.[16] Though this was a more speculative approach to the problem of securing a cargo, the risks were seemingly outweighed by the flexibility and potential reward of being able to purchase on an open market.

Many of the sack ships were based in London, and historians had long assumed that an intense and bitter rivalry existed throughout the seventeenth century between the sack ship interests and the fishing interests, who were based in the ports of England's West Country, particularly those of Devon and Dorset. Only recently has this interpretation given way to a more complex picture, one in which linkages of finance, insurance and business partnerships tied London and the West Country together in arrangements that were complementary more than they were competitive.[17] Nevertheless, the West Country merchants were by far the most consistent and regular participants in both the fishery and the trade, so that Keith Matthews could refer to their role with some justice as a "monopoly."[18] Then came the wars of 1689 to 1713, affecting England, France, Spain, Holland and other European powers. The wars were extremely disruptive, not only of the European markets for Newfoundland fish but also of the transatlantic movement of men and ships which was the definitive characteristic of the migratory ship fishery. The wars in turn were immediately followed by a mysterious failure of the shore fishery in which the cod would not come in. This failure, combined with the effects of the wars, created a depression that lasted well into the 1720s and staggered many of the West Country ports.[19] Some managed to adapt, shifting their capi-

---

[16]See, for instance, Steele, "Instructing the Master," 191-193.

[17]Matthews, "History," 68-73; and Pope, "Adventures," 1-3.

[18]Matthews, "History," 265, as well as his chapter 1, "The Western Adventurers;" Cell, *English Enterprise*, 6; W. Gordon Handcock, *Soe longe as there comes noe women: Origins of English Settlement in Newfoundland* (St. John's, 1989), 53-60; and John Mannion and W. Gordon Handcock, "The 17th Century Fishery," and John Mannion, W. Gordon Handcock and Alan Macpherson, "The Newfoundland Fishery, 18th Century," in R.C. Harris (ed.), *The Historical Atlas of Canada, Vol. I: From the Beginning to 1800* (Toronto, 1988), plates 23 and 25.

[19]Matthews, "History," 307; and Ian K. Steele, *The English Atlantic, 1675-1740: An Exploration of Communication and Community* (Oxford, 1987), 81.

tal into trade or the offshore bank fishery, in which the British at Newfoundland now engaged for the first time.[20] Many, however, reduced their level of participation in the Newfoundland fishery significantly or withdrew altogether. To fill the ensuing vacuum, a number of ports outside the traditional West Country region were now drawn into the Newfoundland trade.[21] There were English ports like Liverpool, Chester and Hull, as well as Irish ports like Cork and Youghal. And, with the appearance of *Christian* at St. John's, there was now a Scottish port, Leith.

The earliest Scottish ship to venture into the Newfoundland trade did so before the end of the sixteenth century.[22] The Scots, however, did not make their presence felt at Newfoundland before the second half of the eighteenth century. David Macmillan pushed the Scots presence at Newfoundland back into the first half of the eighteenth century, but viewed that presence as "a 'cover' not only for the shipping out of colonial produce from the Southern plantation colonies and the West Indies to foreign European ports, but also as a 'blind' for importing large quantities of European goods...into all the British

---

[20]The preference for lightly salted, dry cod in the southern European markets had committed the British fishery almost exclusively to that cure. The French, however, had fished the banks since the mid-1500s, responding to consumer preference in northern France for a heavily salted "green" cod; Christopher Moore, "The Markets for Canadian Cod in the Eighteenth Century: France's Cod Trade and the Problem of Demand" (Unpublished paper presented at the Canadian Historical Association Annual Meeting, University of Guelph, 1984), 4; C. Grant Head, Christopher Moore and Michael Barkham, "The Fishery in Atlantic Commerce," in Harris (ed.), *Historical Atlas of Canada*, I, plate 28. British bank fishermen still brought their heavily salted catch to shore for final curing into the familiar saltfish; Matthews, "History," 311-312; and Head, *Eighteenth Century Newfoundland*, 72-74. Although this hybrid method produced an inferior product, this was more than balanced by the bank fishery's lower overhead, greater productivity (whether measured as an aggregate or per fisherman) and the rising demand of the Caribbean sugar islands for "refuse fish" with which to feed its expanding slave population.

[21]Matthews, "History," 265. Significantly, it was in 1725 that the reference changed in the annual reports on the fishery from "English" sack ships to "British" sack ships.

[22]Matthews fixes 1752 as the earliest year in which there is evidence of a Scottish ship visiting Newfoundland; *ibid.*, 423-424. But the ship *Grace of God* of Dundee made a voyage to Newfoundland and then Lisbon in 1599/1600; W.A. McNeill (ed.), "Papers of a Dundee Shipping Dispute 1600-1604," in Scottish Historical Society, *Miscellany*, X (Edinburgh, 1963), 55-85, esp. 68 and 73.

colonies."[23] There is nothing in Macmillan's discussion to suggest that Scottish trade with Newfoundland was generated, as *Christian*'s voyage so clearly was, by the opportunities provided by the Newfoundland fishery itself. Yet, as we shall see, the men who chartered *Christian* had little if any prior experience with the Newfoundland trade, nor is there any evidence that the voyage of *Christian* was repeated. Instead, that voyage appears to have been something in the nature of a commercial experiment, one prompted at least as much by expectations and conditions in Scotland as by opportunities that seemed to exist in Newfoundland after 1713.

Scotland had only become part of the British North Atlantic commercial sphere with the Act of Union of 1707. For several decades before then, the Scottish economy had struggled within a predominantly European commercial environment that was increasingly mercantilistic and competitive. Scotland's leaders responded with a regime of mercantilistic regulations of their own in the hope of stimulating maritime commerce and economic growth while defending Scottish national sovereignty.[24] The attempt would fail, however. Even under ideal conditions, Scotland's small size made it extremely difficult to improve the country's balance of trade by encouraging exports while actively discouraging imports.[25] When the 1690s ushered in a series of economic crises, the stage was set for what Bruce Lenman has termed "a revolution of frustrated expectations" among Scotland's ruling classes. Negotiations began that eventually led to economic and political union with England in 1707.[26] The immediate benefits of this measure to the Scottish economy included trade protection, access to North American colonial commodity trades and shipping,

---

[23]David Macmillan, "The 'New Men' in Action: Scottish Mercantile and Shipping Operations in the North American Colonies, 1760-1825," in Macmillan (ed.), *Canadian Business History*, 50-51.

[24]Eric Graham, "In Defence of the Scottish Maritime Interest, 1681-1713," *Scottish Historical Review*, LXXI, Nos. 1-2 (1992), 88-109.

[25]Bruce Lenman, *An Economic History of Modern Scotland, 1660-1976* (London, 1977), esp. chap. 3, "Crisis, Union and Reaction, 1690-1727," 45. According to Lenman, Scotland's population stood at roughly 1.25 million in 1695.

[26]Lenman, *Economic History*, 45-52. For a discussion of the economic causes and consequences of Scotland's Union with England, see also Christopher A. Whatley, "Economic Causes and Consequences of the Union of 1707: A Survey," *Scottish Historical Review*, LXVIII, No. 2 (1989), 150-181; T.C. Smout, "The Anglo-Scottish Union of 1707. I: The Economic Background," *Economic History Review*, 2nd ser., XVI, No. 3 (1964), 454-467; and Tom M. Devine, "The Union of 1707 and Scottish Development," *Scottish Economic and Social History*, V (1985), 23-40.

and the revitalization of trade with Iberia and the Mediterranean.[27] Yet Union was not followed by the degree of trade expansion or prosperity that many Scots had expected. Not only did trade now find itself encumbered by a more complex, energetic and burdensome customs and excise establishment, but the limitations of established industries, such as the inferior quality of Scottish linen, were now exposed to the full force of competition. As Tom Devine explains, the Scots would soon learn that "Union provided a series of risks and an array of opportunities. Ultimately what mattered...was the Scottish response to these stimuli."[28] It is in this context that we must examine the voyage of *Christian* to Newfoundland in 1726.

Christian was owned by William Hutton Sr. and his brother Alexander, who also served as the ship's master.[29] They, in turn, had chartered the ship to a limited partnership that included William Hutton and seven others, all of whom, with one exception, were merchants of Edinburgh.[30] At least four of the partners – Hutton, Baillie William Carmichael, Baillie James Newlands and William Jamisone – were linked through domestic investments, including a sugar refinery and distillery and a stagecoach company in Leith. They were also partners in direct trade with Spain both before and after 1726, through which they would have developed commercial contacts in Iberia that would greatly simplify the challenge of branching into the Newfoundland trade.[31] Several of the partners were also prominent members of the Edinburgh and Leith community, having substantial wealth and holding positions of public

---

[27]Graham, "Defense of the Scottish Maritime Interest," 107-108.

[28]Devine, "Union of 1707 and Scottish Development," 32.

[29]SRO, RD, 2d ser., "Durie's Office," 1 December 1736-30 April 1737, "Disposition, William Hutton to spouse Christian Thomson," 28 February 1737. The ship was almost certainly named after William's wife.

[30]The investors included Baillie William Carmichael, Baillie James Newlands, Mr. Robert Dundas, William Hutton Sr., William Jamisone, Walter Scott, and Robert Smith Jr., all of Edinburgh or Leith, and Robert Robertson, a merchant of Glasgow.

[31]SRO, CC 8/8/106, 26 March 1742-20 January 1743, 118-118v, testament of the late William Hutton, merchant in Edinburgh and Residenter in Leith; and SRO, CC 8/8/98, 30 July 1736-23 April 1736, 241-244v, testament of the late Baillie James Newlands. In their letter to their agent in Barcelona, the partners referred to at least one previous commercial voyage to that port; Burd, "Journal," 17v, William Hutton Sr. and partners to William French, 14 May 1726. *Christian* would journey to Spain for salt in 1734; SRO, CC 8/8/106, 118-118v, William Hutton, testament.

trust.[32] Devine maintains that such diversification in both overseas trade and the domestic economy, together with the effective use of a network of kin and close personal acquaintances, was typical of the greater Scottish merchants who sought thereby to "minimize the insecurities of trade, and so...preserve their fortunes."[33] In short, the partners who chartered *Christian* in 1726 were men experienced in manufacturing, domestic and Continental trade, and linked by an informal but advantageous network of social, political and commercial contacts through which a shift into the Newfoundland trade could be made more secure.

Concern for the security of their investment also motivated the partners to hire Edward Burd Jr. As the ship's supercargo, it became his responsibility to attend to the accounts of the cargo and any other commercial affairs of the ship.[34] To have a supercargo on a commercial voyage was no longer as common as had been the case in the seventeenth century, owing in part to the expanding network of overseas agents for British mercantile firms.[35] Nevertheless, as Ralph Davis noted, the supercargo continued to be found in newer trades "where firm business connections had not yet been established."[36]

<hr />

[32]Both Carmichael and Newlands were "baillies" while Hutton, Carmichael, Smith, and Scott were burgesses and guild brethren; Charles B. Boog Watson (ed.), *Roll of Edinburgh Burgesses and Guild-Brethren, 1701-1760* (Edinburgh, 1930). Like his father before him, Hutton was also a member of the Merchant Company of Edinburgh, holding several positions of trust within the Merchant Company and being elected Master of the Merchant Company in 1726, the same year that *Christian* voyaged to Newfoundland; Alexander Heron, *The Rise and Progress of the Company of Merchants of the City of Edinburgh, 1681-1902* (Edinburgh, 1903), appendix V, "Masters of the Merchant Company," 390-392; and Merchant Company of Edinburgh (MCE), minutes, 1681-1696; MCE, register of entrants with the company, 1681-1902; and MCE, minutes, II, 1702-1714.

[33]Tom M. Devine, "The Social Composition of the Business Class in the Larger Scottish Towns, 1680-1740," in Devine and David Dickson (eds.), *Ireland and Scotland, 1600-1850: Parallels and Contrasts in Economic and Social Development* (Edinburgh, 1983), 167; and Devine, "The Merchant Class of the Larger Scottish Towns in the Later Seventeenth and Early Eighteenth Centuries," in George Gordon and Brian Dicks (eds.), *Scottish Urban History* (Aberdeen, 1983), 103-104.

[34]William Falconer, *An Universal Dictionary of the Marine* (London, 1780; reprint, Paderborn, 2012), 284.

[35]Davis, *Rise of the English Shipping Industry*, 123-124.

[36]*Ibid.*, 163. This was the first long-distance voyage for the twenty-four-year-old Burd; National Library of Scotland (NLS), MSS No. 5061, "Genealogical History of the Family of Burd of Ford and Whitehall" (1827), 106. Burd's appointment as supercargo may have been more in the nature of an apprenticeship for young Edward and

Burd's presence therefore strongly suggests that the partners had no previous experience in the Newfoundland trade. This is further indicated by the investors' need to arrange for a letter of credit from Claud Johnson, a merchant in London with good connections in the fishery, to support the bills of exchange that Burd would have to write when purchasing fish in Newfoundland.[37] The voyage of *Christian* can therefore be defined as an experimental investment, possibly motivated by a desire to expand and to capitalize upon the investors' established trade links with Spain and thereby rise into the highest ranks of the Edinburgh merchant community.[38]

"Experimental," however, did not mean that the partners were venturing into the complete unknown. William Hutton, at least, also invested in the Scottish fishing industry and possibly in the European fish trade, for he owned a minority share in a fishing buss and fishing equipment at the time of his death.[39] The Scottish fishing industries had long produced whitefish and herring for export to northern Europe. The control of the German, Dutch and other Continental merchants over that trade was declining by the late seventeenth and early eighteenth century, and Scottish investors were anxious to take over. The conventional wisdom has been that lack of capital, combined with the continued dominance of German merchants in northern Europe, inhibited success, and that lack of experience discouraged significant development

---

a favour to his father, with whom William Hutton had served as Merchant Councillor of Edinburgh years before; Helen Armet (ed.), *Extracts from the Records of the Burgh of Edinburgh, 1701 to 1718* (Edinburgh, 1967), 27, entry for 23 September 1702. T. McAloon views Burd's presence on the *Christian* as "a finishing school for [Burd's] commercial training." See his "A Minor Scottish Merchant in General Trade: the case of Edward Burd 1728-39," in John Butt and J.T. Ward (eds.), *Scottish Themes: Essays in Honour of Professor S.G.E. Lythe* (Edinburgh, 1976), 17.

[37]SRO, RH 15/54/4, A2, letter of credit from Claud Johnson, 28 April 1726. The worth of Johnson's credit was attested by Solomon Merrett, John Spackman, John Lloyd and David Milne, who were later described by the supercargo as "verry well known in Newfoundland but especially Merrett & Hoyte [sic; the original clearly identifies him as Lloyd] are best known in fferryland." See Burd, "Journal," 10v.

[38]Devine identifies participation in foreign trade as the distinguishing criterion between the top two tiers of the merchant community hierarchy in the larger towns of early eighteenth-century Scotland. "Only a minority of all merchants had either the capital or the contacts to pursue this more lucrative type of commerce on a consistent basis." Devine, "Social Composition," 166.

[39]SRO, RD 3/195, disposition, William Hutton to Christian Thomson, 28 February 1737.

of a trade with southern Europe for several decades.[40] But the Act of Union of 1707 was followed almost immediately by petitions to the Board of Trade for subsidies to support a trade in Scottish whitefish to the Mediterranean.[41] And six of the fifteen vessels that *Christian* found at Barcelona in 1727 carried "dryed Cod fish, from Scotland." Indeed, according to Edward Burd Jr., Scottish fish was a preferred product in the Barcelona market, suggesting that Scotland's trade in fish with Iberia was better established by then than has hitherto been suspected.[42] There is also the intriguing fact that during the closing years of the War of the Spanish Succession, several London merchants had employed William Taverner, a Newfoundland ship's captain and trader, to develop a fishery on the northwest coast of Scotland.[43] While none of these points can be linked directly to the voyage of *Christian*, we can assume that William Hutton and his partners had sufficient information at their disposal to encourage an attempt to broaden an existing bilateral commercial pattern with Spain into a triangular pattern that included Newfoundland. Certainly the instructions issued to Captain Hutton and Edward Burd indicate some understanding of the fundamentals of that trade, such as sending *Christian* to St. John's, which by then had established itself as the "Metropolis" of Newfoundland; most ships trading to Newfoundland made it their first or their final port

---

[40]James J.A. Irvine and Ian A. Morrison, "Shetlanders and Fishing: Historical and Geographical Aspects of an Evolving Relationship, Part I," *Northern Studies* XXIV (1987), 44-47; and Hance Smith, *Shetland Life and Trade, 1550-1914* (Edinburgh, 1984), 19 and 22-23.

[41]James D. Marwick (ed.), *Extracts from the Records of the Convention of the Royal Burghs of Scotland, Vol. IV: 1679-1711* (Edinburgh, 1880), "convention approved of several overtures for the encouragement of trade," 25 November 1707, 427-428; and "memorial of several particulars concerning trade, and especially the fishing on the coast of Scotland, 9 April 1711," 511-516.

[42]"Scots fish never fail to Sell here at about 1 p⁸ [ryall] p Quintall more than any fish that come to this place; The reason they give for it is this, that the Scots fish allwayes Stand the Summer better than any other: another advantage they have is, that they soke in double the quantity of Water that any other fish doe, Soe that with 1 lb of fish they sell another of Water." Burd, "Journal," 21. It seems likely that the Scots were selling stockfish, which used no salt in the curing process. A quintal was 112 lbs.

[43]TNA/PRO, CO 194/5/109 (LAC microfilm), petition of William Taverner to Lord Oxford, Lord High Treasurer, n.d., enclosed with William Taverner to the Lords Commissioners for Trade and Plantations, 31 March 1714. Significantly, the letter of credit with which the partners backed the bills of exchange drawn by Burd in Newfoundland was witnessed by a servant of "J. Taverner Sr.," conceivably William's brother. See David B. Quinn, "William Taverner," *Dictionary of Canadian Biography, Vol. III: 1741 to 1770* (Toronto, 1974), 617-620; and Handcock, *Soe longe as there comes noe women*, 47, 50 and table 2.1, 48-49.

of call.[44] The Scots would also have found the relatively low entry requirements for the Newfoundland trade in terms of capital appealing, for they were
as ambitious for success within the new British maritime commercial environment as they were constrained by the limits of their capital.[45] If *Christian's*
voyage to Newfoundland was therefore something of a novelty for William
Hutton and his partners, it was also a practicable and sensible application of
their resources.

III

Nevertheless, as that voyage was to demonstrate, luck was as essential an ingredient as ambition and opportunity for anyone venturing into the Newfoundland fish trade, and this was something which William Hutton and his fellow
investors were singularly lacking. From the start, the voyage of *Christian* was
dogged by misfortune. First, *Christian* took almost twice as long to complete
its voyage as was then the average for West Country sack ships.[46] In part, this
was because *Christian's* point of departure, the east Scottish port of Leith, was
poorly situated for transatlantic journeys. Bad weather also slowed the ship's
progress – almost immediately upon its departure from Leith in the afternoon
of 3 June 1726 the ship was forced to anchor in the Firth of Forth to sit out a
gale; a few days later, another storm drove the ship to seek shelter in Kirston
harbour in the Orkneys, where it remained for the better part of a week owing
to persistent contrary winds.[47] As a result, *Christian* did not clear St. Kilda to
the west of the Outer Hebrides until the end of the second week of its voyage.
Thereafter, variable winds and frequent calms continued to slow the ship's

---

[44]According to the commodore of the Newfoundland station ships in 1715, the
prices given at St. John's "for any thing, is a necessary enquiry for other places to govern themselves by, and the same of their owne Manufactures the price of Fish...and the
whole country enquires how the rates goe at St. John's;" PRO, Admiralty (ADM)
1/2006, I (Captains' Letters), Thomas Kempthorne (*Worcester*, St. John's) to Secretary
Burchett, n.d. [1715].

[45]Gillian Cell remarked on the degree to which "the Newfoundland fishery
[was] run by men of limited capital;" Cell, *English Enterprise*, 6. Devine emphasized
that in the late seventeenth and early eighteenth centuries, the number of Scottish merchants with the necessary finance, experience and contacts to engage in overseas trade
on a consistent basis would have been very small; Devine, "Merchant Class," 98-99.

[46]*Christian* arrived at St. John's over eight weeks after departure from Leith;
the average voyage between the West Country and Newfoundland took about five
weeks, and could take as little as three weeks, according to Steele, *English Atlantic*, 82.

[47]Burd,"Journal," 3 June-30 July 1726.

progress. Yet bad weather and contrary winds alone were not to blame; *Christian* was a slow ship, and a leaky one as well.[48]

The slow passage proved to be critical, for on the very day that *Christian* worked its way into St. John's harbour, the price of fish "broke" at sixteen shillings or thirty-two ryalls per quintal – a full shilling higher than anyone there had expected.[49] Determined to secure a cargo at a lower price, the supercargo, Edward Burd Jr., spent several frustrating days following the advice of Captain Richard Newman, the fishing admiral of St. John's and chasing down elusive rumours of cheap fish.[50] How Burd must have envied Captain Rannie, the Scottish master of a London sack ship, who had arranged to purchase a cargo of bank fish at twenty-seven ryalls per quintal before he left England![51] Finally giving up on "His Ld Ship," as he now referred contemptuously to Captain Newman, Burd made his way first to Bay Bulls, and eventually to Ferryland. There, on 12 August, he finally contracted to purchase a

---

[48]"The Ship proved leaky in the Passage, which obleiged us to heave her down [upon arrival in St. John's] to Search her bottom;" Burd, "Journal," 9-9v, Edward Burd Jr. to William Hutton Sr., 17 September 1726.

[49]30 July 1726: "At Night Capt Weroy in the Crown, a Galley belonging to London, broke price for Shoar fish at 16 S[hillings] or 32 Ryalls p. Quintal, for which he was heartily cursed by allmost every body, even the Boat keepers themselves told that they did not expect above 30 Ryalls." Ordinarily the buyers in St. John's met to settle the price of fish together; this was called "breaking the price." Weroy apparently violated the custom in order to assemble a cargo as quickly as possible, hoping to recoup the higher cost of purchasing the fish by getting to market early enough to command an even better selling price. *Crown* expected to depart Newfoundland on 7 September, a full two weeks before *Christian* began its voyage to Spain; SRO, RH15/54/4, A10, Alexander Wylly to Edward Burd Jr., 5 September 1726; and Burd, "Journal," 22 September 1726.

[50]His search took him first to Torbay, north of St. John's, where the fishermen refused to lower their prices to the level Burd wanted, and then south to Bay Bulls, where two days of negotiations and a promising verbal arrangement fell through; Burd, "Journal," 2-10 August 1726. Captain Newman's status as "fishing admiral" was a custom of the fishery; "the first Ship in the Harbour is Lord or Admirall of that Harbour, & the Captain sits Supream Judge of all differences that happen in the place, from whose sentence there is no appeal but to the Commadore." Burd, "Journal," 13. The fishing admiral also took his choice of fishing room or berth, making this "a worthwhile prize of local control for the first ship there each year." Steele, *English Atlantic*, 83.

[51]Burd, "Journal," 5 August 1726. Later, as *Christian* sailed by the Azores, it spoke to a pink of London, bound for Alicante with a cargo of salt cod from St. Pierre, for which purchase had been arranged before the vessel set out for the fishery at a price of twenty-five ryalls per quintal.

mixed cargo of 800 quintals of bank fish and 700 of saltfish at fourteen shillings six pence per quintal in the mistaken belief that "Bank fish...wou'd doe fully as well as the Shoar for a Spanish Mercate."[52] In return, he was able to sell most of the biscuit brought from Scotland for fourteen shillings per hundredweight; the balance was sold at a reduced price four weeks later.[53]

A few days later, on 22 September, *Christian* set out for Spain, "biding adieue to fferryland & all the Rogues in it," and more uncertain than ever whether the war that threatened between England and Spain would erupt before they reached their destination.[54] Clearly, Burd was not pleased with the venture thus far. Though he had managed to purchase the fish at the price he wanted, it was at a cost of almost two weeks' time, a delay that would prove even more costly later in the voyage. Nor was Burd satisfied at the price he received for the biscuit. While he blamed this on the large numbers of ships that had appeared that season in Newfoundland, "all of them Bringing some [biscuit]," he later conceded that "Bread was sold in the Spring at 20 sh p

---

[52]Burd, "Journal," 11 August 1726. On the acceptability of bank fish, see the entry for 5 August. Yet according to a West Country merchant later in the century, the markets "never take bank fish if they can get shore;" evidence of Mr. Jefferey to the British Parliamentary inquiry, 1792, cited in Matthews, "History," 312. Burd's confidence that bank fish would be accepted in the Spanish market in 1726 signifies either how serious was the collapse of the inshore fishery during the decade after 1713 or else his own inexperience and gullibility.

[53]Burd, "Journal," entries for 12 and 13 August and 11 and 17 September 1726. Burd bought most of the fish from Thomas Holdsworth at Ferryland, though some was received from other individuals as debt payments to Holdsworth; SRO, RH15/54/4, A14. In turn, Holdsworth purchased most of the biscuit – 6000 lbs.; SRO, RH15/54/4, A15, "contract between Thomas Holdsworth & Edward Burd Jr. to purchase bread." Unfortunately, the original purchase price of the biscuit in Scotland is the one detail not found in the otherwise rich body of documents associated with the voyage of *Christian*. Burd was convinced that he would have secured an even better price for the fish, had his bills of exchange been drawn upon credit based in Exeter rather than London ("They will allwayes sell their fish a Ryall a Quintal cheaper for Exeter Bills, than for Bills upon [London]," 22 September 1726). This was a reflection of the fact that, like most Newfoundland communities, Ferryland had developed strong social and commercial links with just one West Country community and its outports, in this case Exeter. Ferryland was a major centre for the bank fishery; according to Burd, it attracted between thirty and forty ships per year, most of them from Limpston near Exeter.

[54]Burd, "Journal," 22 September 1726; and SRO, RH15/54/6, B5, Edward Burd Jr. (Ferryland) to Edward Burd Sr., probably 17 September 1726.

%."[55] Unless a sack ship timed its arrival earlier than the others – and there was little doubt that *Christian* was unfortunate in this regard – it could not hope to sell its cargo at the highest possible price.[56] As for the fabrics and other dry goods that he had hoped to sell on his personal account, here Burd had only marginal success. He had managed to sell a considerable quantity of one consignment, albeit by peddling it in small individual lots to an assortment of individuals, but he only managed to sell a portion of a second, and of a third, none.[57] What passed for communities in Newfoundland at this time were small and rude. Their residents lacked the means to purchase, and the merchants, factors and agents (who brought supplies of their own with them from England) lacked the need to demand in volume the kind of frippery Burd had to sell.[58] As an inhabitant of St. John's later explained to Burd, the commodities most in demand in St. John's were rum, molasses, salt and provisions, none of which were part of *Christian's* manifest.[59]

Burd knew, of course, that these setbacks would be more than balanced by a good price for the fish in Spain and a profitable cargo for the final leg of the journey home. But time, which had already undermined the profits of the voyage thus far, now conspired to ruin it. As *Christian* made its way to

---

[55]Burd, "Journal," 11 September 1726. Burd discusses the prices of various commodities in a lengthy description of the fishery, inhabitants and resources of Newfoundland that appears under the entry for 22 September 1726.

[56]"I expected allwayes that the price of bread would have risen, but Such great numbers of Ships coming & all of them Bringing some, Still kept the price Low;" *Ibid.*, 11 September 1726. Timing was immensely critical to a sack ship, both for the sale of its cargo and the purchase of fish. In response to his request for advice on "what things was wanting in this place," Burd was told that "unless you Could be a Erly Shipe one Dare not venture To advise;" SRO RH15/54/4, A22, Alexander Wylly (St. John's) to Edward Burd Jr., 13 December 1726.

[57]So disappointing were the results of his efforts to peddle the goods on his private account that Burd was compelled to plead with his father for an advance of money to be picked up in Cádiz; SRO, RH15/54/6, B5, Edward Burd Jr. (Ferryland) to Edward Burd Sr., probably 17 September 1726.

[58]According to the 1726 survey of the fishery, the summer populations of St. John's and Ferryland were 436 and 477, respectively; adjusting for seasonal and temporary residents, the number of permanent residents, whose dependency on imported goods was greatest, would have been roughly 264 and 256, respectively – much too small to serve as an adequate consumer market for the number of sack ships that year. On the interpretation of early eighteenth-century Newfoundland population data, see Handcock, *Soe longe as there comes noe women*, 95-98.

[59]SRO, RH15/54/4, A22, Wylly (St. John's) to Edward Burd Jr., 13 December 1726.

the Straits of Gibraltar, it became abundantly clear that it was the wrong ship for the trade. It took two weeks to make it to the Azores, then it took another six days to sail through that mid-Atlantic chain! Seven weeks after departure from Ferryland, the Scottish sack ship experienced the humiliation of being passed by ships that had all departed Newfoundland three weeks behind *Christian*. Arriving at Gibraltar on 15 November, Burd quickly checked with the three or four merchants there with whom it was customary to leave mail for ships, but there was nothing, neither from his employers nor from his employers' factor at Sanlúcar, Mark Pringle. This did not discourage Burd from writing several letters of his own, summarizing the voyage thus far and indicating *Christian*'s intention of setting out immediately for Barcelona, where "ffish ...gives a great price...at 7½ Dollars [per hundredweight]." Honesty, however, forced him to concede that *Christian*'s slow passage from Newfoundland almost certainly would bring them to Barcelona behind the rest of the trade, "tho' wee were the first Ship from Newfoundland bound to our Port."[60] Of the war with Spain that had long been rumoured, Burd heard only mixed reports, causing him some unease. A more immediate concern were the fickle winds that hindered *Christian*'s departure from Gibraltar and dragged its journey to Barcelona out to more than two weeks. There, more bad news awaited them – not only did they discover that much of their cargo of fish had spoiled, but the price of fish in the Barcelona market was lower than expected.

The damage to the fish was considerable. Of the 1500 quintals of fish received in Ferryland, only 217 quintals were judged first quality; over a thousand quintals were second grade, and more than 200 quintals were third grade.[61] On top of that, the fish sold for only six dollars per quintal, not the seven and a half dollars he had anticipated in Gibraltar. It is possible that the poor quality could not have been avoided. Gregory French, the investors' agent in Barcelona, emphasized that "the Same fate Attended all the Ships with fish that arrived this Season att this port," and back in Newfoundland, the commodore indicated in his report to the Board of Trade that the late appearance of the cod that year, combined with a wet summer, had made it difficult to cure the fish properly.[62] Burd's decision to buy a mixed cargo of bank fish and shore fish may also have been a factor in the deterioration of the cargo.

---

[60]Burd's letters of 15 November 1726 to Claud Johnson, William Hutton Sr., and Mark Pringle, all in Burd, "Journal," 16-16v.

[61]*Ibid.*, 18v, "Accompt Sale of 1500 Quintals Poorjack from Newfoundland." A quintal of second-grade fish fetched only two-thirds the price of a quintal of first-grade fish; third-grade fish fetched only half the price of first grade.

[62]SRO, RH15/54/4, A3, Gregory French to William Hutton Sr. and Co., 12 January 1727; and TNA/PRO CO 194/8, 38v, Commodore Bowler, "Answers to the Heads of Inquiry," October 1726.

Yet French blamed condensation in the hold, exacerbated by the length of the voyage and the fact that "the Ship it Selfe was Not extrardy tight." Burd predictably preferred to blame the ship as well, though he was reluctant to say so to his employer.[63] He was convinced that, notwithstanding the damage to the cargo, the fish might have fetched a higher price had *Christian* arrived sooner at Barcelona. Instead, Burd lamented, "No Ship that have come either here or to Alicant have had soe long a passage as Wee." Fifteen ships had arrived there before them, so that the fish buyers had already purchased all the fish they normally required for the year – about 30,000 quintals, according to Burd; by the time *Christian* appeared, "they...were become verry cool in buying."[64]

Adding further injury to the situation, *Christian* now encountered persistent problems in making up its return cargo once the fish had been unloaded.[65] First, there was no cork to be had in Barcelona; the agent, Gregory French, attributed this to an unusually bad season which made cork not only scarce and expensive but also poor in quality. *Christian* therefore took on several small cargoes – some wine for Cádiz, some empty wine pipes for Mataró and more wine for delivery to Gibraltar.[66] French claimed that freighting these small cargoes would be more rewarding for the owners than continuing to wait for cork to appear, a prospect he regarded as "good for Nothing."[67] Yet every additional cargo meant another delay. By the time they departed their last port of call at Villasa, the season for a quick passage to the Straits was over.[68] It took *Christian* a full month to reach Gibraltar, fighting gales and storms all the way. *Christian* also ran afoul of two Algerine cruisers, though

---

[63]"The Ship realy was not in soe good a Condition as She ought to have been; tho' I have Writt nothing of this to Mr Hutton." SRO, RH15/54/6, B1, Edward Burd Jr. (Barcelona) to his father, n.d. [probably 31 December 1726].

[64]Burd, "Journal," 19v, Burd to William Hutton Sr., 31 December 1726. See also Burd's lengthy commentary on Barcelona, 20v-23, esp. 21.

[65]Unloading the fish took from 8 December to 22 December, with frequent delays occasioned by a combination of bad weather and numerous "holy dayes." Burd, "Journal."

[66]SRO, RH15/54/6, B54, charter party between Richard Neiland (Barcelona) and Alexander Hutton, 8 January 1727.

[67]SRO, RH15/54/4, A3, Gregory French to William Hutton Sr. and Co., 12 January 1727. Burd could not have been pleased to learn subsequently that within days of *Christian*'s departure for Gibraltar, French had "more Cork than Youl be able to take in," SRO, RH15/54/6, B53, French to Edward Burd Jr., 23 January 1727.

[68]Vilasar de Mar, a tiny port north of Barcelona and just south of Mataró.

they delayed the Scottish merchantman only long enough to extort some fish, meat and wine.[69] When they finally arrived at their destination, it was to learn that the long anticipated conflict between England and Spain had begun at last. Gibraltar was besieged, and all trade was immobilized until convoys could be arranged for their protection.[70]

For Burd, the sojourn in Gibraltar appears to have been something of an entertaining diversion. He spent much of his time there peddling the wares left in his private consignment, shopping, making new friends, studying Gibraltar's commercial possibilities and recording his impressions of the Spanish siege as it unfolded.[71] For Captain Hutton, however, the enforced idleness, combined with the rapid disintegration of the freighters' original arrangements for the voyage – the intended stop in Sanlúcar for wine and cork was now clearly out of the question – became increasingly unbearable. At one point, and against both his instructions and his supercargo's advice, Hutton became determined to proceed to Leith in ballast.[72] But Hutton would not sail without the protection of convoy, and besides, the winds were unfavourable, so that *Christian* remained at Gibraltar until the middle of March. By then, Hutton and Burd had patched up their differences and agreed to sail in ballast in accordance with their original instructions as far as Bordeaux, taking their chances with the Spanish privateers in the Bay of Biscay in order to pick up a cargo of wine and salt. When they arrived there on 4 April, the two men parted company. Hutton headed for the island of St. Martin-de-Ré, off La Rochelle, to load salt before departing on 12 May for the final homeward journey; Burd dallied a while in Bordeaux before returning to Scotland at a leisurely pace by way of Paris, Boulogne and London.[73]

---

[69]Burd, "Journal," 24 January 1727.

[70]*Ibid.*, entries, 14 January-14 February 1727.

[71]This part of the journal has been transcribed by R.K. Hannay and published as "Gibraltar in 1727," *Scottish Historical Review*, XVI, No. 4 (1919), 325-334.

[72]Burd, "Journal," 27 February 1727. Their original instructions directed them to proceed in ballast to Bordeaux for new orders should they not be able to secure freight in Barcelona or elsewhere in the Mediterranean. Burd, however, wrote in his journal that, "Mr Hutton...told me that he designed to goe for Leith North about, I Represented to him the Inconveniences that would attend it, he told me it was none of my bussiness, & that he was resolved to goe home." See also SRO, RH15/54/6, B50, receipt, Alexander Hutton to Edward Burd, 27 February 1727.

[73]SRO, RH15/54/6, B7, Edward Burd Jr. to Edward Burd Sr., 6 June 1727. To their surprise, Hutton and Burd found no instructions awaiting them in Bordeaux. In taking on salt at St. Martin's, Hutton therefore acted either on his own judgment or in accordance with his knowledge of his brother's needs.

At this point, Burd's journal and letters lose touch with *Christian*. We can assume, however, that the journey home from St. Martin-de-Ré went no more quickly than any other leg of its voyage and that the ship would have been lucky to arrive home within a year of its departure from Leith on 3 June 1726. Indeed, for *Christian* to be lucky at that stage of the voyage would have been something of a reversal of fortune, since bad luck seemed to dog it at every turn – the contrary weather as it attempted to clear Scotland and later as it struggled to reach Gibraltar from Barcelona; the arrival in St. John's on the day that the price of fish broke against their favour; what was possibly a bad season for properly cured fish; the unusual lack of cork in Barcelona; the outbreak of war with Spain just as the ship reached Gibraltar, effectively immobilizing it there for a month. Yet much of this bad luck could also be attributed to the poor sailing quality and condition of the ship; it was slow and leaky to the detriment of its speed and its cargo in a trade where the quality of both were crucial. As a result, it seems unlikely that the investors profited by the voyage. Without an indication of how much they paid for the biscuit they shipped to Newfoundland, it is impossible to judge whether a profit was made on that leg of the voyage, but the poor quality of the fish and the low price when the ship reached Barcelona almost certainly meant that any profit on the second leg of the voyage was insignificant.[74] The small freights carried between Barcelona and Gibraltar would hardly have covered the expenses of sitting idle for a month in Gibraltar or travelling from there to Bordeaux in ballast. The final leg, with salt from Bordeaux to Leith, would not have earned a tremendous profit.

IV

In the decades immediately following Scotland's Union with England in 1707, Scottish merchants were keen to take advantage of the new economic opportunities that they assumed would ensue from participation in British colonial trade. At the same time, the English West Country domination of the Newfoundland fishery and trade was momentarily shaken by the stress of many years of a war that only ended in 1713, followed immediately by the stress of a depression within the fishery. It seemed an auspicious moment for Scots merchants to venture into the Newfoundland trade. With a modest investment, some familiarity with the domestic Scottish fishing industry and trade, and some experience in direct trade with Spain, William Hutton of Leith and several partners chartered the ship *Christian* to voyage as a sack ship to New-

---

[74]Ignoring for a moment various sundry charges, commissions, etc., the fish cost slightly under £1100 in Ferryland and sold for about £1142 in Barcelona (5832 Ryalls, two sols at an exchange rate of forty-seven pence per ryall); Burd, "Journal," 18v and 26v, 26 February 1727.

foundland and Barcelona. That the voyage appears to have been unprofitable does not necessarily make it a failure. It has been the contention of this paper that the voyage was more in the nature of a commercial experiment, an attempt to assess whether an established trade with Spain could be made more profitable by transforming what, to that point, had been a bilateral trading pattern into a triangular one. The lessons suggested by the voyage of *Christian* seemed to be that it could not. Blame it on bad luck, on the wrong ship, on locational disadvantages, or on the hazards of an unfamiliar trade – William Hutton and his partners appear to have concluded that the time had not yet come for Scots merchants to become involved in the trade with Newfoundland.

"*Une petite Republique*" in Southwestern Newfoundland: The Limits of Imperial Authority in a Remote Maritime Environment[1]

Introduction

This paper continues earlier research into the origins and persistence of settlement by French and Irish fishermen at Codroy in southwestern Newfoundland from about 1725 until 1755.[2] A previous paper identified the settlers as French fishermen whose lineage could be traced back two generations to fishing families living on the French coast of Newfoundland during the 1680s who were forced to abandon the island following the Treaty of Utrecht (1713) by which France acknowledged British sovereignty over the entire island. A subsequent paper identified one of the metropolitan French merchant families whose capital had maintained the earlier fishing stations in Newfoundland and who reappeared as traders and investors at the Codroy settlement during the 1730s and 1740s. This paper integrates the results of those earlier studies and extends the analysis beyond 1744 when the outbreak of war caused the abandonment of the Codroy settlement. The community was re-established in 1748, only to be destroyed a second time when hostilities between France and England resumed in 1755. When Codroy reappeared in 1762, it was as part of the British mercantile world, for the principal merchant was English. Though the social and economic character of the community after 1762 resembled the earlier pattern, the "French connection" had been broken. This history of a small fishing settlement, in which two French phases were followed by an English one, suggests certain conclusions about the principal factors governing the social and economic character of an isolated community of "people of the sea." I argue that

[1]This essay appeared originally in Lewis R. Fischer and Walter Minchinton (eds.), *People of the Northern Seas* (St. John's, 1992), 1-33. This paper is based in part on research undertaken with support from the Social Sciences and Humanities Research Council of Canada. It was first presented at the Sixth Triennial Conference of the Association for the History of the Northern Seas, Kotka, Finland, August 1992.

[2]"'Une Grande Liaison:' French Fishermen from Île Royale on the Coast of Southwestern Newfoundland, 1714-1766 – A Preliminary Survey," *Newfoundland Studies*, III, No. 2 (1987), 183-200; "'Bretons...sans scrupule:' The Family Chenu of Saint-Malo and the Illicit Trade in English Cod During the Middle of the 18th Century," in *Proceedings of the Fifteenth Meeting of the French Colonial Historical Society Martinique and Guadeloupe, May 1989* (Lanham, MD, 1992), 189-200.

national differences were relatively insignificant, for the similarities between the earlier and later communities at Codroy were stronger than the disparities. The paper is based largely on manuscript and cartographic evidence gathered in the national and regional archives of littoral England and France.

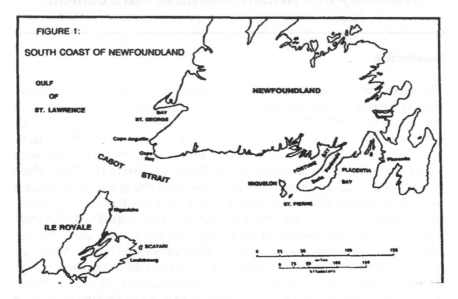

Figure 1: The South Coast of Newfoundland

Source: Courtesy of the author.

The Persistence of French Settlement in Newfoundland (c. 1684-1744)

On 25 September 1714, the departure of *Héros* from the little town of Plaisance in southern Newfoundland marked the end of an era. On board was Philippe Pastour de Costebelle, who had been governor of the French colony since 1695. In the negotiations that ended the War of the Spanish Succession in 1713, France had agreed to cede its fishing colony at Plaisance together with all claims to Newfoundland, even though she had been undefeated on the island.[3] To avoid disrupting the French fishery, which had resumed as soon as hostilities had ended and before the final terms of the Treaty of Utrecht had been defined, England allowed the French to delay their withdrawal from

---

[3]Jean-François Brière, "Pêche et politique à Terre-Neuve au XVIIIe siècle: la France véritable gagnante du traité d'Utrecht?" *Canadian Historical Review*, LXIV, No. 2 (1983), 168-187; and James K. Hiller, "Utrecht Revisited: The Origins of French Fishing Rights in Newfoundland Waters," *Newfoundland Studies*, VII, No. 1 (1991), 23-39.

Newfoundland until 1714. Nevertheless, most inhabitants of Plaisance, the neighbouring settlements and the islands of St. Pierre were already evacuated to Cape Breton by the time *Héros* sailed; Costebelle was one of the last to leave. Henceforth, all Newfoundland, Plaisance included, would belong to the English.[4] After fifty years of effort to develop a viable cornerstone of empire based on the fishery, France had given up on Newfoundland.

It seemed a logical decision. The island's limited resources had forced the Plaisance colonists into what one historian described as "a dangerously single-minded concentration on the fishery." This, in turn, had led them into a nearly total dependency on the merchants of the metropolitan fishery for provisions, supplies and labour. The demands of the fishery, together with a chronic friction between resident and migratory fishermen for beach space, had encouraged a dispersed settlement pattern. While this served the inshore fishery well, it undermined both the centralizing efforts and the authority of government. In the end, Plaisance may have been more a liability than an asset of the French mercantile empire.[5] Turning to Cape Breton Island as a *tabula rasa*, France would try once more to establish its linchpin of empire. French hopes were based on the expectation that Île Royale (as the new colony was known) offered more fertile soil and would attract a significant number of Acadians away from the Bay of Fundy who would give the new colony the economic diversity that had been missing in Newfoundland; the transplanted Placentians were expected to form the nucleus of a new, more successful fishing community.[6]

Yet if France had abandoned Newfoundland, the same cannot be said for all the French inhabitant fishermen. Of a total population of roughly two hundred, perhaps fifty or sixty chose to remain when the English arrived to take control of Plaisance (or Placentia, as it now became known). A small handful lived in Plaisance itself. Most, however, resided in small outposts and fishing stations scattered around Placentia Bay or even further to the west,

---

[4]Georges Cerbelaud Salagnac, "Philippe Pastour de Costebelle," in *Dictionary of Canadian Biography, II: 1701-1740* (Toronto, 1969), 509-513 (*DCB*).

[5]John Humphreys, *Plaisance: Problems of Settlement at this Newfoundland Outpost of New France, 1660-1690* (Ottawa, 1970); and B.A. Balcom, *The Cod Fishery of Isle Royale, 1713-1758* (Ottawa, 1984), 14. For a discussion of the rationale behind the efforts of the French crown to develop Plaisance, see Laurier Turgeon, "Colbert et la pêche française à Terre-Neuve," in Roland Mousnier (ed.), *Un Nouveau Colbert: Actes du Colloque pour le tricentenaire de la mort de Colbert* (Paris, 1985), 255-268.

[6]The relatively small population of Plaisance severely limited the role its transplanted inhabitants could play as the "nucleus" of a new fishing society. Most of the fishermen who eventually settled in Île Royale would come from Europe; John Robert McNeill, *Atlantic Empires of France and Spain: Louisbourg and Havana, 1700-1763* (Chapel Hill, NC, 1985), 18.

beyond the Burin Peninsula at St. Pierre and in neighbouring Fortune and
Hermitage Bays, at Fortune, Grand Bank, Havre Bertrand, Connaigre, Courbin and Hermitage (see figure 2).[7] These tiny settlements had not only been
physically remote from the administrative centre at Plaisance but also had developed a social and economic dependency on St. Pierre that minimized their
need for contact with the main colony. The settlements rarely consisted of
more than one or two nuclear families. The *engagés* who worked for them
were temporary migrants whose presence swelled the total number to nearly
nine times that figure.[8] A memorial submitted by merchants of Saint-Malo
some time before 1713 claimed that *"il y a à la Baye de Fortune et lieux cir-
convoisins, trois ou quatre habitations qui peuvent avoir 50 a 60 hommes. Ces*

---

[7]Excluding the garrison and the indentured fishermen who were but temporary residents, John Humphreys estimates a permanent population that averaged twenty
to thirty male inhabitants in the years between 1671 and 1706, together with more than
one hundred dependent women and children; Humphreys, *Plaisance*, 7. He excludes,
however, the inhabitants of outlying fishing stations such as Petite Plaisance, Pointe-
Verte and Fortune in his calculations, residents who are very much central to this paper. Humphreys used data taken from census lists in France, Archives nationales, archives d'outre-mer (ANO), G^1/467; these were published under the title "Recensements
de Terreneuve et Plaisance" by Fernand-D. Thibodeau in *Mémoires de la société géné-
alogique canadienne française*, X, Nos. 3-4 (1959), 179-188; XI, Nos. 1-2 (1960), 69-
85; XIII, No. 10 (1962), 204-208; and XIII, No. 12 (1982), 245-255. In using these
census data, Humphreys warned that the "vague terminology and differing methods of
enumeration preclude any precise calculation;" *Plaisance*, 22n. This contributes to a
large discrepancy in estimates of the numbers of people who remained behind at Placentia after the evacuation. In *Atlantic Empires*, 18, John McNeill claims there were
"fewer than ten," while Jean-Pierre Proulx suggests a larger figure of fifty to sixty in
"The Military History of Placentia: A Study of the French Fortifications" in his *Placentia, Newfoundland* (Ottawa, 1979), 118-119. This is consistent with subsequent
English records that included the more remote settlements of Fortune and Hermitage
bays.

[8]Researchers of the early settlement history of Newfoundland are careful to
distinguish between seasonal, temporary and permanent migrations to the island. The
permanent population was much smaller than the total figures in seventeenth- and eighteenth-century censuses might suggest. See John J. Mannion (ed.), *The Peopling of
Newfoundland: Essays in Historical Geography* (St. John's, 1977), 5; Keith Matthews,
"17th Century Settlement in Newfoundland" (unpublished paper, 1974), 5; and W.
Gordon Handcock, *Soe longe as there comes noe women: Origins of English Settlement
in Newfoundland* (St. John's, 1989), 40. Handcock suggests that an accurate measurement of the permanent population of Newfoundland would be to double the number of
women and add the children (2F + C), on the grounds that "the number of women and
children may be regarded as an index of the more stable and permanent population,
since it is axiomatic that these categories have implications for the germination, per-
petuation and continuity of a population and its social capacity to absorb subsequent
immigrants (intermarriage)." Handcock, *Soe longe as there comes noe women*, 95.

habitans font la pesche tout l'année lorsque le tems est beau et le permet."[9]
The earliest recorded census for the region, taken in 1687, recorded an overall
population of perhaps forty adults and children (see table 1).[10]

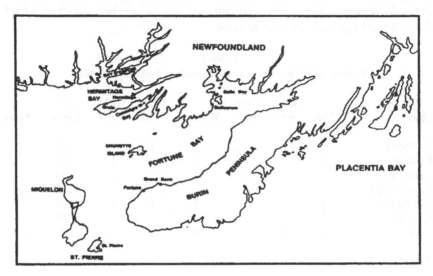

Figure 2: St. Pierre and Dependencies

Source: Courtesy of the author.

Despite their small numbers, the inhabitants of Fortune and Hermitage bays were permanent residents of Newfoundland. Although the contract
labourers were for the most part temporary, some of the families had been
living in these tiny communities for a generation by the time Plaisance was
evacuated in 1714. Gilles Vincent and his wife, Marguerite Durand, were
identified as residents of St. Pierre in 1691, 1693 and 1694; in 1714, he had
moved, but only as far as Connaigre. Laurens Millou, his wife Marie LeMan-
quet, and their children were at Fortune in 1693, 1694 and 1711; in 1714, they
were at St. Pierre. Julien and Noel Petit were both at Hermitage in 1694; in
1711, Julien was still at Hermitage, but Noel was at Fortune, while in 1714,
he was reported to be living at Connaigre. One of the most permanently rooted
settlers was Jean Bourny, who lived with his wife and children at Grand Bank

---

[9]Cited in Jean-François Brière, *La pêche française en Amérique du nord au
XVIIIe siècle* (Saint-Laurent, PQ, 1990), 63n.

[10]ANO G¹/467, Nos. 4 and 5.

in 1691, 1693, 1694, 1711 and 1714.[11] Most families were also related by marriage. Bourny's wife, Marie Commaire, was probably related to three other women named Commaire (Anne Marie, Simonne and Jeanne) who were living with their husbands at St. Pierre in 1693 and 1694. The names Legrand, Durand, Vincent, Commaire and Bourny would all become intertwined through marriage in the decades to come.

Table 1
Population of St. Pierre and Dependent Settlements, 1687

Place	Men	Boys +15	Boys -15	Women	Girls +15	Girls -15	En-gagés	Total
St. Pierre	3	3		3	1		66	76
Fortune	1	1	3	1	2	1	63	72
Grand Banc	3			2		1	39	45
Havre Bertrand	1						35	36
Cap Negre	5	1	2	3	1	1	58	71
L'ermitage	2			1			28	31
Total	15	5	5	10	4	3	289	331

Source: See text.

This pattern, in which dispersed settlements consisted of one or two resident families with temporary contract labourers and seasonal fishermen swelling their numbers, was typical of seventeenth-century Newfoundland, both French and English.[12] If there was geographic mobility, residents did not move far; some lived at Fortune or Connaigre one year and at St. Pierre an-

---

[11]*Ibid.*, Nos. 6, 9 and 11; Great Britain, National Archives (TNA/PRO), Colonial Office (CO) 194/6, 243, William Taverner, "Survey of the Inhabitants," 1714; and CO 194/6, 226-241, Second Report for 1714/1715 (received 20 May 1718). To simplify the citation of the French census material and the Taverner survey, they will be cited in subsequent references simply as ANO Plaisance Census, with the year, and PRO Taverner Survey. Unless specifically indicated to the contrary, the dates on English documents are given in the Julian or Old Style, except that the year is taken to begin on 1 January, not 25 March as was then customary. This puts the dates of English documents eleven days behind those of French documents, which are given in the Gregorian style.

[12]Of the thirty or so English settlements in Newfoundland towards the end of the 1690s, eight contained only one nuclear family and twenty-four had less than five families. The English population fluctuated around 2000; Handcock, *Soe longe as there comes noe women*, 39-46.

other.[13] They supported themselves by fishing in the summer and hunting and trapping in the winter. They marketed their catch not through Plaisance but through St. Pierre, which served as both a social and economic centre. Most of the fifteen dwellings reported at St. Pierre in 1690 would have been seasonal abodes for French merchants and traders, especially from Saint-Malo. Their metropolitan connections provided the region's inhabitants with essential services such as outfitting, provisioning and marketing.[14] Both during and after the French era, St. Pierre was "a considerable Place of Trade;" as the summer ended, and "especially about Mich[mas] [29 September]...all the Planters & Servants from the Bay de Espère, Capnigro, Grand Bank, Fortune, Courbin &c. bring in their Furrs and Summers Fish to sell for purchasing their Winters Provisions and necessarys."[15] The merchants also brought *engagés* into the area. Some were contracted to the inhabitants; others remained in the merchant's employ, in which case the resident fisherman may have acted more as an agent than as his own master. A few merchants maintained their own fishing stations. Thus, the proprietor named Boismoiris who maintained a station at Fortune in 1714 was almost certainly Claude Chenu, Sieur de Boismoiry, a merchant of Saint-Malo who with his three brothers played a substantial role in sustaining the resident French fishing communities in southwestern Newfoundland during the 1730s and 1740s.[16] People would also have been encouraged to settle in these remote places by the presence at St. Pierre of a religious cleric, and another at "Cap Negre" (Connaigre).[17]

The disruption and hardship of the war years from 1702 to 1713 must have been difficult. When some of the buildings at Hermitage were "Burnt by the English" in 1710, at least one resident moved into Bay d'Espoir.[18] Never-

---

[13]In 1714, at Bandalore, deep in Fortune Bay, there were several structures that belonged to "Monsr Belorm, a Malouin Gentleman, who hath winterd in that place 20 Years, Successively one after the other;" TNA/PRO, CO 194/6, 232, Taverner, "Second Report of his Survey Work, 1714/1715" (received 20 May 1718). This was most probably Simon Belorm or Bolorm, identified in 1714 as a resident of St. Pierre.

[14]C. Grant Head, *Eighteenth Century Newfoundland: A Geographer's Perspective* (Ottawa, 1976), 13.

[15]TNA/PRO, CO 194/5, 260v, Taverner to Secretary of State, 22 October 1714.

[16]A full treatment of the Chenu family and its activities is provided in Janzen, "'Bretons...sans scrupule.'"

[17]ANO, G[1]/467, Nos. 4 and 5.

[18]TNA/PRO, CO 194/5, 260, Taverner to Secretary of State, 22 October 1714.

theless, by the time William Taverner carried out the first English census of
the region in 1714, the adult population stood at forty-three, with forty-nine
children, for a total of ninety-two; the number of *engagés* was 228 (see table
2). The number of structures had also increased (including at least seven chap-
els).[19] The people were clearly thriving. And, as the people made quite plain,
they were prepared to remain despite the fact that England exercised sovereignty. Taverner reported that most of the male inhabitants and slightly fewer
of the *engagés* had sworn the oath of allegiance to the English crown. In effect, the people had been transformed into Newfoundlanders, and the tiny settlements in St. Pierre and Fortune and Hermitage Bays had become homes
they preferred not to abandon.[20]

This, however, was much more easily said than done. The authorities
at Île Royale were determined to make a success of the new colony and sent a
priest among the French Newfoundlanders to exhort them to leave.[21] Others
were driven away by the British authorities at Placentia who were quick to
abuse their authority, extorting illegal fees and payments from the inhabitants
and confiscating their properties for resale to friends and associates. Between
1714 and the late 1720s, this abuse was so persistent that it drove away not
only the French but also a significant number of new English settlers.[22] A
more subtle yet ultimately more pressing concern was the difficulty of the
French inhabitants in adapting to the new mercantile framework after 1714.
Long accustomed to exchanging their fish with French merchants for essential
supplies and gear, often with the help of credit established over many years of
trade, it was not easy to develop similar commercial relations with British
merchants. For a while, French trading ships continued to stop at St. Pierre.

---

[19]TNA/PRO, CO 194/6, 243, Taverner, "Survey of the Inhabitants," 1714;
and CO 194/6, 226-241, Taverner, "Second Report of his Survey Work for
1714/1715" (received 20 May 1718). The improved ratio of inhabitants to *engagés* was
a logical reflection of the way war interrupted the annual movement of seasonal labour.

[20]One factor which contributed to the resiliency of the permanent population
was the effort made by the residents to grow more of their own food. At the "Bay of
Cap nigro" (Connaigre), Taverner found three plantations which, in addition to its fishery, maintained "2 Com ffeilds in which growes very good Barley, as ever I saw."
TNA/PRO, CO 194/6, 232v, Taverner, "Second Report of his Survey Work for
1714/1715" (received 20 May 1718).

[21]TNA/PRO, CO 194/6, 48-48v and 50, Taverner, "Some Remarks," items F
and M, February 1715.

[22]Head, *Eighteenth Century Newfoundland*, 59-60; and TNA/PRO, CO 194/8,
148v and 150, Capt. St. Lo to Board of Trade, 5 March 1728.

Table 2
Population of St. Pierre and Dependent Settlements, 1714

Place	Name of Planter	Men	Women	Children	Total	Boats	Stages	Houses
Hermitage	Jonathan Micholl	28	–	–	29	6	2	3
Isle Grole	Peter Caroy	6	2	–	9	2	2	5
Canoger	Giles Vincent	17	3	–	21	4	2	4
	Noell Petit	8	–	–	9	2	1	3
	Jonathan Durant	–	–	–	1	–	–	2
St. Peters	Garrn Tulon	19	–	–	20	4	2	6
	Giles Lossout	24	1	–	26	4	1	4
	Simon Bolorm	25	1	6	33	4	1	5
	Larrano Miloro	–	–	–	1	–	1	1
Grand Bank	Jno Vilden	17	2	1	21	4	1	4
	Steph Dcroch	5	1	6	13	1	1	5
	M. Desallones	21	1	11	34	5	1	4
	Jno Bourny	4	1	7	13	1	1	2
	Jno Durand	–	–	–	1	–	–	4
	Julin Durrand	–	–	–	1	–	–	2
	Mr. Mountain	–	–	–	1	–	–	4
	Peter Dupon	–	–	–	1	–	–	2
Fortune	Reno Troquy	9	–	–	10	2	2	2
	Tho. Commes	8	1	6	16	2	1	3
	Jno Muslow	5	1	–	7	–	–	1
	Nic Boaux	7	1	3	12	3	1	2
	Boismoiris	8	–	–	9	–	–	3
	Larr. Million	17	1	9	28	9	2	6
	Jno Manhoy	–	1	–	2	–	1	4

Place	Name of Planter	Men	Women	Children	Total	Boats	Stages	Houses
Goads	Bossumo	-	-		1	-	1	4
	Jno Durant		17	-	1			
Total	26 Planters	228	17	49	320	53	25	85

Source: See text.

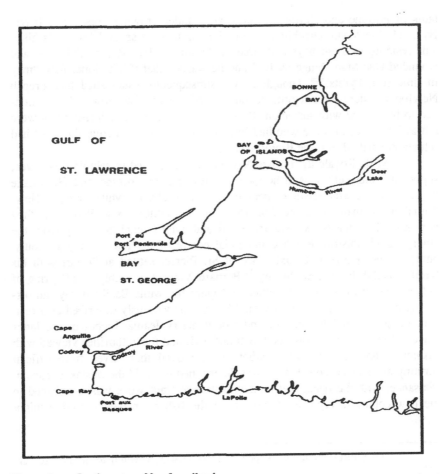

Figure 3:   Southwestern Newfoundland

Source: Courtesy of the author.

This, however, was something the British would not tolerate, and soon there were no more reports of French shipping at St. Pierre or Fortune Bay.[23] Claude Chenu of Saint-Malo made every effort to maintain his fishing station at Fortune, including the extremely expedient, if not very circumspect measure of swearing the oath of allegiance to the English crown. Yet as commercial contact with France came to an end, both he and the French inhabi-

---

[23]TNA/PRO, CO 194/5, 197, Elisha Dobree to Taverner, 5 March 1714; CO 194/5, 117-117v, Moses Jacqueau to William Lownde (Treasury), 7 May 1714; CO 194/5, 260v, Taverner, "Report," 22 October 1714; and CO 194/6, 49v, Taverner, "Some Remarks," item I, February 1715.

tants of Fortune and Hermitage Bays were under growing pressure to leave. By 1724, Lieutenant Gledhill was complaining that the several French families still residing in these bays had fired on "Some of His Majties Subjects, and wounded One Man Dangerously," and he warned that if left alone, they "may in time turn Pyrats."[24] Though it was subsequently established that French Newfoundlanders were legitimate subjects of the English crown, they continued to be viewed with suspicion. By the late 1720s, no French fishermen were left at Placentia; of those who had lived beyond the Burin Peninsula, most had reluctantly drifted to Cape Breton Island.

At Île Royale the former Newfoundlanders tried to rebuild their lives, supporting themselves as fishermen as they had in Fortune and Hermitage bays. As was characteristic of people who had preferred while living in Newfoundland to maintain some distance from the authorities at Plaisance, they now settled in places like Scatari Island and even Niganiche, away from the emerging administrative and commercial centre of the new colony at Louisbourg. Thus, Pierre LeGrand, once of St. Pierre, settled at Scatari with his family, as did the widow Bourny.[25] Bernard Vincent, possibly an offspring of Gilles Vincent of St. Pierre, settled at Niganiche, while Basil Bourny, an offspring of the Bourny family of Grand Bank, was variously described as a resident of Scatari and Niganiche, and would marry Jeanne Pichot of the latter place.[26] But Île Royale was not Newfoundland, and the authorities viewed with suspicion those people who deliberately removed themselves from official scrutiny and supervision. France was determined to avoid the mistakes made at Plaisance, with the result that an administrative framework was established in the new colony to regulate all aspects of the fishery in which friction might

---

[24]TNA/PRO, Admiralty Papers (ADM) 1/1473, Gledhill to St. Lo, 16 July 1724.

[25]ANO G¹/466, Nos. 68 and 69, general census of inhabitants at Île Royale (hereafter cited as ANO Île Royale census, with date), 1726 and 1734. The widow Bourney, described as a native of Plaisance, could only have been Marie Commaire, wife of Jean Bourny; there were no Bournys in French Newfoundland except the family at Grand Bank.

[26]*Ibid.*; Canada, Sessional Paper No. 18, "Journal and Census of Île Royale, prepared by le Sieur de la Roque under the direction of M. le comte de Raymond, in the year 1752," *Report Concerning Canadian Archives for the Year 1905* (Ottawa, 1906), II, 44 (hereafter cited as de la Roque Census); ANO G²/197, Greffes des tribunaux de Louisbourg et du Canada, doss. 153; and G²/196, doss. 124, f. 41. Bourny is described in de la Roque's census as a native of Plaisance, born in 1702. Again, this almost certainly made him a son of Jean Bourny and Marie Commaire for there was no other family by that name living in French Newfoundland at the time.

develop between residents and the seasonal fishermen from France.[27] This strategy quickly became restrictive and burdensome to the point that a substantial number of fishermen began to drift to a place even more remote, free of control and supervision. They began to relocate to the southwestern corner of Newfoundland where they settled in several small harbours. The names "Port aux Basques," "Cape Ray," and "Codroy" were used at various times. These names may have been used interchangeably; the name "Codroy" is itself a corruption of Cape Ray (Ca' de Ray). More likely, they referred to individual communities scattered about southwestern Newfoundland (see figure 3). In any case, this choice of locations placed the inhabitants far enough away from Placentia that they could reasonably expect to avoid interference from the English, who lacked both the will and the means to exercise their authority that far west, and it was effectively beyond the reach of the officials at Louisbourg, whose jurisdiction, according to the terms of the Treaty of Utrecht, did not extend to the coast of Newfoundland. Thus, the French Newfoundlanders of Fortune and Hermitage bays found themselves returning home.

At first, their return was strictly seasonal. In 1724, the Louisbourg authorities accused Basil Bourny, Julien Durand and a fellow named Sabot of maintaining a fishing station at Cape Ray.[28] Their choice of Cape Ray and Port aux Basques was dictated in part by the metropolitan French merchants who employed them. Some of these merchants were Basques who were encountering difficulties and discrimination in securing beach space on Île Royale for their fishing operations and who, like the French Newfoundlanders, gravitated to Niganiche and even Newfoundland in search of a less regulated environment.[29] Certainly that region of Newfoundland had long been associated with the fishing vessels of St. Jean de Luz and Bayonne. The Basques began maintaining semi-permanent shore stations at Codroy in the mid-1720s, leaving crews to overwinter, then picking them up the following year.[30] Other ships

---

[27]Balcom, *Cod Fishery*, 14-15.

[28]AN, Archives des colonies (AC), série C^{11}B/7, 71-71v, Jacques-Ange Le Normant de Mézy, *commissaire ordonnateur* at Louisbourg, to Jean-Frédéric-Phélypeaux, Comte de Maurepas, Minister of Marine, 27 November 1724. Julian Durand, another "native of Plaisance," would marry Madeleine Durand, a native of Niganiche and possibly a granddaughter of Gilles Vincent of St. Pierre; de la Roque Census, 43. Sabot was probably Jean Sabot, husband of Jeanne Bourny and therefore Bourny's brother-in-law.

[29]Laurier Turgeon, "Pêches basques en Atlantique nord (XVIIe-XVIIIe siècle): étude d'économie maritime" (Unpublished doctorat de 3e cycle, Université de Bordeaux III, 1982), 45-46.

[30]TNA/PRO, CO 194/6, 237v, Taverner, "Second Report" (received 20 May 1718); Rather than extend his survey to the west coast himself, Taverner relied on in-

showing up at Cape Ray were from Granville.[31] But it was the merchants of Saint-Malo who were most strongly linked with the French Newfoundlanders, and one merchant in particular: Claude Chenu. He was identified by Father Marcellin, the missionary at Scatari, as the man who recruited local residents to fish for him in Newfoundland and who, despite the proscriptions of the Louisbourg officials, had encouraged Sabot, Bourny and Durand to work seasonally at his fishing station at Cape Ray.[32]

Claude Chenu, Sieur Boismory (c. 1678-17?), was one of four brothers: Pierre, Sieur Dubourg (c. 1683-1769); Jacques, Sieur Duchenot (c. 1687-1758); and Louis, Sieur Duclos (c. 1694-1774). A fifth Chenu, Jerôme, Sieur Dupré (c. 1698-17?) may have been a cousin.[33] They did not belong to the top

---

formation provided by the master of a ship belonging to Saint Jean de Luz. The 150-ton *St. Michel*, built and owned by the Sieur Michel de Pedesclaux of Saint Jean de Luz, made fishing voyages to Newfoundland in 1723 and 1724, on each occasion leaving two shallops with their crews to overwinter at Cape Ray; see Pau, Archives départementales des Pyrenees-Atlantiques (AP-A), série B (supplement), 8724, Registres de l'Amirauté de Bayonne, 1722-1725, entries for 12 January 1724 and 9 August 1725. In 1755, a Basque captain stated that he and his fellow Basques had maintained a shore station at Codroy for thirty years; Bayonne, Archives de la chambre de commerce de Bayonne (ACCB), série 12 (pêche de la morue), No. 40, statement of Jean Lafreche, 9 October 1755.

[31]Brest, Archives maritimes au port de Brest (AM Brest), série 1P¹/1, 1726, 7/14 and 1727, 5/85, ref. to *Thomas d'Aquin*, departed Granville in February 1725 and headed to Cape Ray, Newfoundland.

[32]Archives of Fortress of Louisbourg National Historic Site (AFL), AC C¹¹B/6, 193-198v, Joseph Monbeton de Brouillan, dit Saint-Ovide, Governor of Louisbourg to Maurepas, 24 November 1723; AN, AC B/47(ii), 1260-1275, Maurepas to Saint-Ovide, 26 June 1724; AN, AC C¹¹B/7, 71-71v, de Normant de Mézy to Maurepas, 27 November 1724; AFL, AC C¹¹B/7, 72, affidavit of Rousseau de Souvigny, 23 September 1724; and AFL, AC B/47(ii), 1260-1275, Maurepas to Saint-Ovide, 25 July 1725. Since the pagination of the AFL copies of French archival materials does not always accord with the originals, AFL and AN records are distinguished here.

[33]Information on the family was compiled from various registers, reports, declarations and other manuscripts housed in the Archives départementales de l'Ille-et-Vilaine, Rennes (AD I-et-V), the Archives de l'arrondissement maritime de Rochefort, (AM Rochefort), the Archives départementales de la Charente-Maritime, La Rochelle (AD C-M), as well as archival sources already cited. Also of great use was L'Abbe Paul Paris-Jallobert, *Anciens Registres Paroissiaux de Bretagne (Baptêmes – Mariages – Sepultures): Saint-Malo-de-Phily; Evêché de Saint-Malo – Baronnie de Lohéac – Sénéchaussée de Rennes* (Rennes, 1902); and Paris-Jallobert, *Anciens Registres Paroissiaux de Bretagne (Baptêmes – Mariages – Sepultures): Saint-Malo (Evêché – Seigneurie commune – Sénéchaussée de Dinan)* (Rennes, 1898). The parents of the Chenu brothers were Jean Chenu and Margueritte Porée.

rank of Saint-Malo's *négociants*; they lacked the wealth, diversity of commercial activity and international associations which were definitive characteristics of the great merchants.[34] Instead, their business dealings appear to have been confined to the fishing industry and related activities – outfitting, shipowning and trade. All appear to have spent their younger adult years as captains of fishing or trading vessels which they either owned individually or in partnership with each other. This in itself was fairly typical of all levels of merchants engaged in overseas commerce, though Louis Chenu seemed content to remain a captain-owner throughout much of the 1730s and 1740s even as his brothers were establishing themselves as merchants of Saint-Malo or its suburb, Saint-Servan. Claude Chenu appears to have been the most mobile. At various times he was identified as belonging to Granville, Saint-Malo and La Rochelle, although he always maintained both the business and personal sides of his relationship with his brothers. Together, the Chenus maintained fishing stations, owned vessels, and participated in the truck trade at Île Royale. Their sons provided a large labour pool from which were drawn the captains and junior officers of the Chenu vessels – and the next generation of merchants. This intricate network of shared investment and vessel ownership was fairly characteristic of *"une société familiale,"* by which family connections were used to secure French businesses against the many risks of eighteenth-century commerce.[35] In short, they were a fairly typical, if not yet very powerful merchant family "on the make," hoping perhaps to duplicate the success of others, like Noël Danycan, who had risen from modest means to the highest ranks of Malouin commercial society.[36]

It was precisely for this reason that the Chenus became intimately involved with the fishing stations in southwestern Newfoundland. Indeed, they played a formative role in their transformation from fishing stations to permanent settlements. The men Claude Chenu recruited to work at Cape Ray were ones he had known, perhaps even employed, at his fishing station in Fortune

---

[34]André Lespagnol, *Histoire de Saint-Malo et du pays malouin* (Toulouse, 1984), 152-153.

[35]The importance of family in eighteenth-century French business is discussed in André Lespagnol, "Une dynastie marchande malouine: les Picot de Clos-Rivière," Société d'Histoire et d'Archéologie de l'Arrondissement de Saint-Malo, *Annales* (1985), 233; and Laurier Turgeon, "Les échanges franco-canadiens: Bayonne, les ports basques, et Louisbourg, Île Royale (1713-1758)" (Unpublished mémoire de maîtrise, Université de Pau, 1977), 86.

[36]Lespagnol, *Histoire de Saint-Malo*, 131-132. Noël Danycan, Sieur de l'Epine (1656-1734), was the son of a small outfitter who began his career as a captain-outfitter of a Newfoundland fishing vessel. Worth only 15,000 *livres* at the time of his marriage in 1685, he was worth millions twenty years later thanks to investments in the South Sea trade.

ten years earlier. During the 1720s, Claude and his brother Pierre frequently visited Scatari Island as owner-captains of vessels engaged in the fish trade, while maintaining a station at Codroy Island near Cape Ray. Brothers Louis and Jacques played junior roles, often serving as second captains. In the early 1730s, Jacques assumed a more prominent part in family trade and even came to dominate it. He owned and outfitted many of the vessels; went frequently to Île Royale; and was the most active Chenu in Louisbourg commerce. Pierre increasingly became rooted as a shipowner, outfitter and merchant at St. Servan, while Claude was based in part at La Rochelle, handling affairs there; their vessels often made for La Rochelle during their homeward voyage, picking up salt for the fishery. Louis remained a captain-owner and was a frequent visitor at Codroy and Cape Ray by the late 1730s and the early 1740s. Presumably he supervised family affairs there.

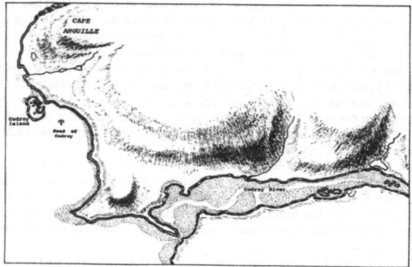

Figure 4:   Codroy Island

Source: Courtesy of the author.

The exact location of the Chenu fishing station is not certain. The documents suggest that a few people lived at Cape Ray but that most resided at Port aux Basques and Codroy Island (see figure 4). The island is small, barren and windswept, bleak in appearance and quite boggy in places; as a place of settlement, the more hospitable and fertile Codroy River valley just a few miles to the south seemed to make more sense. But Codroy Island was settled by fishermen attracted by a different kind of logic, one governed by the excellence of the local cod fishery and the particular qualities of the island that made it ideal for curing fish. According to James Cook, on its landward side was "a small snugg Harbour for Fishing Shallops, wherein is 12 or 14 Feet at high Water...in this Harbour are...convenient places [for stages] with good Beaches for drying of Fish. In the Road of Cod Roy is very good Anchorage for Shipping." William Taverner claimed that the harbour provided good fishing for salmon and cod from April until June, while wood and wild geese were abundant nearby.[37] With easy access to additional salmon fishing, hunting, and good stands of timber in the nearby Codroy Valley, Codroy Island was a sensible place for fishermen to settle.

By the mid- to late 1720s, Codroy had become more than just a seasonal fishing station supported in part by the Chenus. Women appeared early in the 1720s when Codroy was still a seasonal station. Jeanne Bourny, the sister of Basil and daughter of Jean Bourney and Marie Commaire of Grand Bank, must have spent considerable time there because several of the children by her first husband, Jean Sabot, were born at Cape Ray: Antoine (born around 1723), Guillemette or Gilette (born about 1725) and Michelle (born about 1728).[38] By 1734, there were at least ten families living there.[39] Six years later, Louis Colas, a chaplain on a fishing vessel that put in at Port aux Basques, recorded the names of some forty individuals in the course of performing a number of baptisms and marriages; the people, wrote Colas, had not seen a priest in seven years. Nearly two dozen more names were recorded in 1741 when a visiting Recollet priest performed some more marriages and baptisms at Codroy Island.[40] Some of the witnesses belonged to the fishing and trading vessels then in port; one was Louis Chenu. Most, however, were permanent inhabitants, living in multi-generational family units. Many were for-

---

[37]Taunton, Admiralty Hydrographic Department (AHD), C.54/5, James Cook, "A Chart of the Sea-Coast...between Cape Anguille and the Harbour of Great Jervis..." (1766). As early as 1714, William Taverner remarked that "the biscayers generally fish here, but their Ships ride under an Island, about two Leag' to the Soward of the North Cape, where they generally anchor close to the maine." The "North Cape" was Cape Anguille, and the island was Codroy; see Captain Duhaldy's information in TNA/PRO, CO 194/6, 240, Taverner, "Second Report."

[38]De la Roque Census, 66.

[39]TNA/PRO, CO 194/9, 259-259v, especially Nos. 60-62, Lord Muskerry, "Answers to Heads of Inquiry," 1734.

[40]ANO, G¹/410, No. 12, Louis Colas, *"prestre au monier du navire Le Mars," "Extrait de registre de bapteme et de mariage de la poste du Petit Nord nommé Port au Basque,"* 18 May 1740; and ANO, G¹/407, registre 1, fol. 76, "Baptemes fait à lisle de Cadray," 12 July 1741. Mass baptisms and marriages of this type were not uncommon in eighteenth-century Newfoundland where appearances by the clergy were infrequent. See Handcock, *Soe longe as there comes noe women*, 153.

mer French Newfoundlanders who had drifted there from Île Royale: names like Bourny, Commaire, Durand and Vincent were conspicuous.[41] Living with them, and frequently related by marriage, were French fishermen who had either been *engagés* at Île Royale or else arrived with the fishing vessels of Saint-Malo, Granville and Bayonne. There was Jean Nicholas de Malvilain of Saint-Malo; several of his children by his marriage to Marie Magdelaine Durand were baptized at Port aux Basques in 1740. There was Guillaume le Marechal, a fisherman from Carolle on France's Cotentin Peninsula; the daughter by his marriage to Jeanne Sabot was similarly baptized at Port aux Basques. A significant proportion were not even French. About half were Irish who may have drifted west from Placentia, where a substantial influx of Irish labourers had occurred in the 1720s.[42]

The people supported themselves principally through the fishery, turning to trapping in the winter. They exchanged the fish and furs for supplies and provisions from passing ships. The Cabot Strait was a gateway for French shipping bound for Canada, the fishery of western Newfoundland, the Gulf of St. Lawrence and the "Grand Bay" towards Labrador. Vessels put in at Port aux Basques and Codroy either to fish, to trade for fish from the inhabitants or to trade with one another. It was this trade that kept the settlements alive. According to Taverner:

> ...their greatest Supply is from St. Malo and Rachael [La Rochelle], who Supply them with almost every thing Needfull, to carry on the Cod Fishery, Salmon-Fishery, Seal-Fishery, and Furring, nay even with green Men, which are Engaged for thirty Six Months, which the Masters in France pretends...Serve at Cape Breton.[43]

---

[41]In 1740, Louis Colas performed a marriage between Jeanne Bourny, now widowed, and Pierre Berteau. Pierre was a native of French Newfoundland; his father was also Pierre Berteau and his mother was Renée Carmel. In 1752, Renée was still alive, though she was by then 102 years old and living with Pierre and Jeanne in Ance Daranbourg in Île Royale; de la Roque Census, 65-66. Renée was the sister of Marie Carmel, who witnessed some of the marriages and baptisms at Cape Ray in 1740. At the same time that he married Pierre and Jeanne, Colas baptized Pierre, their first son, who was born in September 1739. Jeanne's brother, Charles, married Anne Vincent.

[42]Several children belonging to Daniel Rourque and his wife Anne Macnamara and to Francis O'Neill and Georgina Hill were baptized at Port aux Basques in 1740. Magdelaine Hill was married to Louis Thebaut; three of their children were baptized at Codroy in 1741. The son of Augustine Hill and Hélène Porré was baptized at Codroy in 1741. See ANO, G¹/407, registre 1, fol. 76; and ANO, G¹/410.

[43]TNA/PRO, CO 194/23, 180-182v, Taverner to Board of Trade, 2 February 1734.

A French Minister of Marine later complained that two vessels from Saint-Malo and Granville "*a donné du Secourse a des deserteurs françois et anglois qui se refugies au port des basques situé prée de l'Isle Royalle ayant fourny...des hommes et des Ustenciles malgré les deffensce faites tant par le Gouverneur de L'Isle Royale.*"[44] Clearly, the people of Port aux Basques and Codroy would not have survived had it not been for regular visits by metropolitan ships and vessels.

It was a situation quite characteristic of eighteenth-century Newfoundland, where "merchants...were the activating agents" of settlement.[45] In this particular instance, however, the activities of the merchants were neither approved nor desired by the authorities at Louisbourg. From the start, officials were concerned by the way in which the Newfoundland settlements allowed the fishermen and inhabitants of Île Royale to abandon ("*abandonner*") Île Royale.[46] As the settlements persisted, they acted as temptations to the *engagés* employed in Île Royale to desert their employers and their contracts, in the process of which some also stole their employers' shallops and gear.[47] Saint-Ovide referred to Port aux Basques as a hideout or lair ("*ce repaire*") for those evading the law, and its inhabitants were variously described as "*deserteurs*" and "*fripons*" at best, and "brigands" and "bandits" at worst. What most disturbed officials was that in Newfoundland, unlike Île Royale, the fishermen were free of restrictive and burdensome regulations, fees, and supervision. As Bigot succinctly explained, "*c'est qu'ils y sont independans.*" Another official

---

[44]AM Brest, 1P¹/2, 1738, 10/60, Maurepas to M. de la Fossingnant, *commissaire-général de la marine* at Saint-Malo, 4 March 1738. The vessels were *Le Mars* (Saint-Malo, 200 tons, 112 men), and Captain Jean Baptiste Le Pelley, Sieur des Cerisiez, in concert with *Le Vandangeur* (Granville, eighty tons) Captain Olivier Grentel. *Le Mars* was also the ship on which the priest, Louis Colas, served when he visited Port aux Basques in 1740.

[45]Gordon Handcock, "English Migration to Newfoundland," in Mannion (ed.), *Peopling of Newfoundland*, 24. John J. Mannion, "Settlers and Traders in Western Newfoundland," *ibid.*, 234, maintains that in Newfoundland, "Unlike so many parts of frontier North America, established trading patterns preceded permanent settlement."

[46]AN, AC B/47(ii), 1260-1275, Maurepas to Saint-Ovide, 26 June 1724.

[47]AN, AC B/52-2, 607v-608, Maurepas to Le Normant de Mézy: "*...nombre de matelots et engagez y ont enlevé cette plusieurs chaloupes à leurs maistres et les ont emmené en l'Isle de terreneuve dont quelques unes ont esté reprises.*"

characterized the society of Port aux Basques and Codroy as *"cette petite Re-publique."*[48]

Of possibly equal or greater concern was the way in which the Cape Ray settlements developed into a centre of illicit trade, for not all the merchants who traded at Port aux Basques or Codroy were French. In 1732, Governor Saint-Ovide reported that Anglo-American as well as French vessels were supplying the people at Cape Ray with supplies and provisions in exchange for fish and oil; Anglo-American vessels were found at Port aux Basques as early as the late 1720s.[49] For several years, Jeanne Bourny's brother-in-law by her second marriage, Antoine Berteau, managed a fishing station at Port aux Basques for an unidentified English merchant.[50] In 1734, William Taverner complained that a French planter at Port aux Basques with the improbable name of Russell had died two years earlier still indebted to him, and "tho' the Man Dyed worth Money," the other residents conspired to prevent Taverner from collecting his debt.[51] What drew the New Englanders was not only the fishery and trade but also the opportunity to trade with French merchants without fear of supervision or interference by the authorities.[52] Yet these commercial activities were ultimately responsible for the continued survival of the Cape Ray communities. The French authorities recognized this fact and would use it in their attempts to bring settlement in southwestern Newfoundland to an end.

The ideal solution would have been to evict the inhabitants from Newfoundland altogether. Tempting though the idea was, France exercised no jurisdiction in Newfoundland because of the Treaty of Utrecht.[53] Officials there-

---

[48]AN, AC C^{11}B/24, 17v, Bigot to Maurepas, 4 October 1742; and TNA/PRO, State Papers (SP) 78/207, 675, memorial enclosed in J. Burnaby (Paris) to John Courand, 2 March 1735 (N.S.).

[49]AFL, AC C^{11}B/12, 254-2, 62, Saint-Ovide to Maurepas, 14 November 1732. In 1727, a Boston vessel at Port aux Basques was attacked and seized by Micmac Indians from Cape Breton Island; AN, AC C^{11}B/9, 50v-51 and 64-70v, Saint-Ovide to Maurepas, 13 September and 30 November 1727.

[50]De la Roque Census, 44.

[51]TNA/PRO, CO 194/23, 180-182v, Taverner to Board of Trade, 2 February 1734.

[52]AFL, AC B/75, 84-84v, Maurepas to de la Bove, 27 April 1742.

[53]Saint-Ovide transmitted to Maurepas the suggestion of the British commander at Canso, Nova Scotia, to launch a joint Anglo-French expedition against Cape Ray. No immediate action was taken, Maurepas preferring to handle the question at that point through diplomatic channels; AFL, AC C^{11}B/12, 254-262, Saint-Ovide to Maure-

fore tried to discourage the movement of people to Newfoundland at its source in Île Royale. Numerous orders and proclamations were issued, demanding that the fishermen at Cape Ray return at once to French territory. At one point, officials even directed the priest at Niganiche to stop performing marriages between women of that outport and men from Cape Ray out of recognition that the viability of the communities in Newfoundland depended less on the number of men who settled there than the number of women who joined them.[54] This, too, proved futile; in the absence of an effective enforcement mechanism, orders forbidding migration were difficult, if not impossible, to enforce.[55] Officials at Louisbourg then tried to interdict trade with Newfoundland in the realization that commercial linkages with France and America were nurturing the settlements.[56] This strategy proved no more successful than the others – a confirmation, perhaps, that regulation and supervision of overseas commerce and possessions were more theoretical ideals than practical measures in the eighteenth century.

The fact was that French authorities lacked the means to investigate merchants suspected of illegal trade. There were few warships stationed at Louisbourg, and these were expected to patrol the fishing banks, not adjacent coastlines or in search of illicit trade.[57] To monitor and regulate trade, Paris

---

pas, 14 November 1732. France eventually offered to organize such an expedition if the British were to grant permission; TNA/PRO, SP 78/207, 675, memorial enclosed in Burnaby to Courand, 2 March 1735 (N.S.). By then, however, British concerns that the settlements at Codroy and Port aux Basques were French and therefore in violation of the Treaty of Utrecht had eased following a visit to Port aux Basques by Captain Crawford in HMS *Roebuck*. Crawford assured his superiors that the settlers had all sworn allegiance to England and were subjects of the crown; TNA/PRO CO 194/9, 259-259v, Lord Muskerry, Answers to Heads of Inquiry, Nos. 60-62, 1734. Any thought of allowing the French to evict British subjects from British soil was quickly dismissed.

[54]AN, AC C¹¹B/24, 117-117v, Bigot to Maurepas, 4 October 1742.

[55]See, for instance, the proclamation of Governor Saint-Ovide to the *"habitans françois du port aux basques,"* AD, C-M B-274, 28 June 1734.

[56]In 1733, the French government ordered a *"Défense absolue"* on all French merchantmen trading with Cape Ray and declared its intention to punish those who had been doing so. Yet ships continued to go there. See AFL, AC C¹¹B/12, 254-262, Saint-Ovide to Maurepas, 14 November 1732; AFL, AC B/59-2, 522-523v, Maurepas to Saint-Ovide and Le Normant de Mézy, 19 May 1733; and AFL AC C¹¹B/14, 43-50v, Saint-Ovide and Le Normant de Mézy to Maurepas, 13 October 1733.

[57]McNeill, *Atlantic Empires*, 92-93, 239n. F.J. Thorpe concedes that while Louisbourg was expected to play a significant role in Maurepas' efforts to restore the French navy after 1723, naval growth failed to keep up with the naval infrastructure established at Louisbourg; "The Cod Fishery in French American Strategy, 1660-1783"

relied on port officials in Normandy, Brittany and the Basque country. Yet to judge by the ineffectual efforts to stop trade with Newfoundland, this approach was simply not reliable. Louisbourg officials seemed to expect their counterparts in France to crack down on illicit trade with Newfoundland, while port officials like the *commissaire-général de la marine* at Saint-Malo, de la Fossingnant, seemed to perceive the problem more as a colonial than a metropolitan one. In any case, whatever their feelings about the affront to their authority that the *"petite Republique"* might represent, officials at Louisbourg or Saint-Malo adopted a fairly tolerant attitude towards illicit trade if it meant prosperity for their respective ports.[58] For this reason, by 1742 the *ordonnateur* at Louisbourg was claiming that more harm than good would ensue should the communities at Codroy and Port aux Basques be eliminated.[59] Consequently, the French authorities seemingly could do little more than suspect and accuse merchants of illicit trade, to which the merchants responded either with excuses or assurances that any illicit activity would cease – but of course, it did not.

The family Chenu provided a striking demonstration of this practice. The men were flagrant participants in the trade at Codroy; certainly they attracted considerable attention, not to mention threats, from the authorities both in Louisbourg and Paris. Yet at no time was there any evidence that the Chenus came into serious difficulties because of their association with fishing settlements. Nor did the Chenus try very hard to disguise their activities. Cape Ray appeared in the port records of Saint-Malo as the intended destination of two of Jacques Chenu's vessels in 1743 and 1744, while Codroy was the intended destination of another in 1743; in that same year, Claude Chenu was co-owner of a vessel destined for Cape Anguille, which looms over Codroy Island. Indeed, at one point they not only admitted that their ships were destined for Cape Ray but also argued that they should be permitted to maintain a

---

(Unpublished paper presented at the Annual Meeting of the Canadian Historical Association, Québec, June 1989), 5-6.

[58]On illicit trade between Louisbourg and New England, Andrew Hill Clark, *Acadia: The Geography of Early Nova Scotia to 1760* (Madison, WI, 1968), 319, remarked that "Not only did the Louisbourg officials wink at it, but at times they encouraged it, both because of the real need of the colony and because its extra-legal nature made it possible for the officials to profit from it."

[59]AFL, AC C¹¹B/24, 28-30, Governor Duquesnel and *ordonnateur* Bigot to Maurepas, 17 October 1742. By 1742, the fishery at Île Royale was experiencing a fall in production; Bigot wanted permission to allow French ships to make up their cargoes with purchases of New England cod if necessary. Additional urgency was provided by the war between England and Spain; French merchants were scrambling to satisfy Spain's enormous demand for cod with the Spanish markets closed to English fish. See Janzen, "Bretons...sans scrupule," 193.

fishing station on the grounds that the fishery there was extremely productive; they responded to charges of trading at Cape Ray by insisting that any trade was with other French fishing vessels.[60] Nevertheless, there is much evidence that their activities in southwestern Newfoundland were being covered up. Louis Chenu spent some time at Codroy with his vessel, *Légère*, in 1741; Chenu was both a godparent and a witness that July when a visiting Recollet priest performed some marriages and baptisms. Yet the voyage summary recorded upon his return to Saint-Malo in December gave no indication that there had been any ports of call between his departure from Saint-Malo on 3 June and his arrival at Laurembec on Île Royale an otherwise unaccountable two months later. Other Chenu vessels made equally lengthy and suspicious ocean crossings but cannot so easily be proven to have put in at one of the ports of southwestern Newfoundland.[61]

By 1743, even Maurepas was sceptical that trade with the Cape Ray settlements could successfully be interdicted, though he never gave up entirely on the idea. Success would come only with the cooperation and assistance of the English, who alone exercised sovereignty and therefore jurisdiction over southwestern Newfoundland.[62] The English, however, could never be roused to act vigorously against the settlements, even though the Board of Trade had been emphatic that "it was not for the Interest of the Fishery of Newfoundland to encourage Settlements there, even of His Majesty's Subjects."[63] This was not an age in which government was noted either for its familiarity with specific colonial possessions or for the vigour of its administration of empire.[64] When government could be roused to investigate reports of settlers at Port aux Basques or Codroy, officials were soon satisfied that neither British sover-

---

[60]AM Brest, IP¹/2, 1733, 19/67, Maurepas to M. de 1a Fossingnant (Saint-Malo), 31 March 1733.

[61]*Hirondelle* (or *Irondelle*): AD I-et-V, 9B/501, 13 March 1743, and 9B/420, 157, 21 March 1744; *Achille*: AD I-et-V, 9B/171, 7 June 1743; and *Saint-Ursin*: AD I-et-V, 9B/171 and 9B/420, 22 May 1743. For *Légère*, compare AD I-et-V, 9B/501, 5 December 1741 with ANO, G¹/407, registre 1, fol. 76, "Baptemes fait à lisle de Cadray 1741."

[62]AFL, AC B/76, 508-509v, Maurepas to Governor Duquesnel and Bigot, 30 June 1743.

[63]TNA/PRO, CO 194/23, 176, Board of Trade to Duke of Newcastle, 24 April 1734.

[64]James Henretta, *"Salutary Neglect:" Colonial Administration under the Duke of Newcastle* (Princeton, 1976), 31-33, 64-67, 104-105, 107 and 266-267; and Ronald Hyam, "Imperial Interests and the Peace of Paris (1763)," in Hyam and Ged Martin (eds.), *Reappraisals in British Imperial History* (Toronto, 1975), 21-43.

eignty nor the British fishery at Newfoundland were in jeopardy. It is ironic, then, that the first phase of settlement at Codroy came to an end not because of government action, but through its inaction. When England and France went to war in 1744, the tiny fishing settlements in southwestern Newfoundland were utterly defenceless; the very ambiguity of authority that had left Louisbourg officials frustrated by their inability to bring these settlements to an end, and which had left the British indifferent to their presence, now left the inhabitants without protection against marauding privateers. Choosing discretion over valour, the residents, both French and Irish, withdrew to the seeming security of Île Royale. Within a few months, their homes and properties had been destroyed by privateers.[65]

The Resumption of French Settlement in Newfoundland (c. 1748-1755)

The inhabitants who fled their homes in southwestern Newfoundland remained true to the independence of character which Bigot had found so lamentable. Those who can be identified by name tended to return to the outports of Île Royale, especially Scatari, Niganiche and Petit Bras d'Or. Some went to Louisbourg; of these, a few refused to join the general evacuation that followed the capture of the town in 1745. When the war ended in 1748, most appear to have remained in their new homes rather than return to Newfoundland. Yet Codroy, Port aux Basques and the other settlements in southwestern Newfoundland were re-established by the early 1750s. Who settled there and in what numbers has not yet been determined; perhaps the question will never be answered. Yet by the time the next war flared in 1755, the population at Codroy and its environs was substantial and gave every indication of once again filling the kind of role it had played in the 1730s and 1740s.

What little we know about the rebirth of these settlements comes from documents describing their destruction in 1755. Though France and England were not officially at war, their armies had already clashed deep within North America. In Nova Scotia, British civil, military and naval authorities met to discuss what to do with the several thousand French-speaking Acadians who had been living under their jurisdiction since 1713 but had consistently refused to swear an oath of allegiance to the British crown. As a precaution against their giving assistance to France in the fast-approaching war, the authorities agreed to the controversial solution of expelling the entire Acadian population from the province. Lost in the enormity of this decision is the fact that those same authorities also decided to remove the hundred or so people living in southwestern Newfoundland. Upon his arrival in Nova Scotia, Edward Boscawen, Vice-Admiral of the Blue, learned that French fishermen had settled

---

[65]AFL AC C¹¹B/26, 32-36, Governor Duchambon and Bigot to Maurepas, 4 November 1744.

once again at Port aux Basques.[66] As preparations for the Acadian deportation were begun, he ordered three ships to Newfoundland to deal with the situation.[67]

Commanding this expedition was John Rous, in the *Success* frigate; he was accompanied by *Arundel*, Captain Thomas Hankerson, and *Vulture*. Rous was an obvious choice for the job. As a New England privateering captain, he made extensive raids on French fishing fleets and harbours in the Petit Nord of Newfoundland in 1744. In July 1755, he participated in the decision to expel the Acadians from Nova Scotia; in October, he would be in charge of the convoy that removed the Acadians of the Chignecto Isthmus.[68] In effect, the operation in Newfoundland was a small dress rehearsal for the greater tragedy in Nova Scotia.

In company with *Arundel* and *Vulture*, Rous sailed from Halifax early in August; after a brief stop at St. Peter's he arrived at "Cape Ray Road" (Codroy Road) on 21 August. While the other ships went off to cruise the adjacent trade lanes and to reconnoitre Port aux Basques, Rous proceeded to seize the Basque vessels he found at Codroy, together with their fish, oil and other marine products.[69] Assisting him were four English inhabitants of Port aux Basques (though whether they had been coerced is not clear). All the inhabitants at Codroy this time appear to have been French; many fled in shallops while one small vessel managed to slip its moorings before it could be seized. Nevertheless, at least sixty-seven inhabitants, mostly women and chil-

---

[66]W.A.B. Douglas, "Edward Boscawen," in *DCB*, III, 70-71.

[67]TNA/PRO ADM 1/481, 67-69v, Boscawen to Admiralty, 15 November 1755.

[68]W.A.B. Douglas, "John Rous," in *DCB*, III, 572-574.

[69]TNA/PRO, ADM 51/940, ii, captain's log, *Success*, Captain Rous; and TNA/PRO, ADM 52/1092, iii, master's log, *Vulture*, Mr. Joseph Marriott. The vessels seized were *Ste. Catherine* (150 tons) of Saint-Jean-de-Luz, Captain Sansin Halsouet, and *Deux Maries* (140 tons), Captain Jean Lafreche. *Vulture* later sailed north to Bay St. George where it took *Saint-Jean*, Captain Marticot Hiriboure (*Vulture*'s log identifies it as *St. Jean Baptiste*); to judge by the destruction ashore of "Tents & Craft," this was definitely a seasonal camp with no hint of permanence. Both *Saint-Jean* and *Deux Maries* were owned by Esteban Dargainarats of the Basque port of Ciboure. See statements of the captains of the fishing ships in ACCB I.2, Nos. 40, 41, 44, 46, 47 and 48. All three ships had departed Socoa, near Ciboure, in February *"pour aller au Cap de Ray en terre neuve faire la Pescherie et Sescherie de Morues."* They had arrived early in April and had just begun loading the fish that their crews had made that season when they were seized. *Notre Dame du Rozaire* of Saint-Malo managed to escape, but not without leaving sails, new cables, an anchor, several shallops, some fish and *"Tous les Coffres des equipages"* behind.

dren, were taken. The crews of the Basque vessels were caught almost entirely unaware by this aggression and watched helplessly as their catches were seized and the shore facilities, including cabins, stages, and shallops, put to the torch. Five harbours in total were devastated, their facilities and habitations destroyed and four vessels taken (including at least two merchantmen captured while making their way from Bordeaux to destinations in French Canada). Rous then delivered his prisoners to Île Royale where they were deposited unceremoniously on the beach at Gabarus Bay.

What little we can learn from these events about settlement in southwestern Newfoundland between 1748 and 1755 is instructive. Once again, several harbours, most notably Port aux Basques and Codroy, had attracted not only seasonal residents but also, to judge by the number of women and children, permanent inhabitants. Once again, the existence of these settled outposts depended not simply on the fishery but on metropolitan merchants and traders who maintained fishing stations and presumably included residents among their employees. There is a suggestion that Port aux Basques harboured more English-speaking residents while Codroy was predominantly French. Though we cannot attach a single name to any of these residents, the speed with which the communities were re-established after peace was restored in 1748 suggests strongly that their inhabitants were the same people who had abandoned them in 1744 and were simply returning to their Newfoundland homes. Finally, and also once again, the remoteness which enabled the little society at Codroy to escape official notice and allowed it to grow in peacetime was its undoing in time of war. The second phase of French inhabitancy in southwestern Newfoundland had come to an even more abrupt and traumatic end than the first. As it turned out, 1755 also marked the end of the last phase of French residence. When Codroy again reappeared in the documentary record early in the 1760s, it would be as a British settlement.

The Final Phase of Inhabitancy (c. 1760-1783)

When the Seven Years' War came to an end in 1763, Newfoundland station ships began patrolling the west coast for the first time. In 1764, Governor Palliser became the first governor of Newfoundland to visit the area, bringing most of his ships in a deliberate assertion of British sovereignty. In 1766 and 1767, James Cook surveyed the coast for the British Admiralty. These various official visits were responses to French claims after 1763 that the fishing rights and privileges they enjoyed under the terms of the Treaty of Utrecht included the entire west coast of Newfoundland.[70] The British Board of Trade had quickly challenged this claim with some cartographic research, adding that

---

[70]Olaf Janzen, "Showing the Flag: Hugh Palliser in Western Newfoundland, 1764," *The Northern Mariner/Le Marin du Nord* III, No. 3 (July 1993), 3-14.

"many of Your Majesty's Subjects have long been used to frequent and carry on a fishery in those Harbours, particularly that near Cape Ray and in the Bay of St. George."[71] This was indeed stretching the truth, for most of the "Subjects" of King George III in the region had been French until 1755.[72] It was to add weight to these conclusions that Hugh Palliser was directed to "visit all the Coasts and Harbours of the said Islands and Territories under Your Governmt," a directive which took Palliser and most of his ships to the Bay of Islands during the summer of 1764.[73] It was also largely for this reason that the authorities offered no objections and took no actions against the British merchants who began to nurture settlement on that coast, even though elsewhere in Newfoundland it was official policy to regard permanent settlement as a threat to the well-being of the migratory fishery and to discourage its spread.

When Captain Samuel Thompson of the warship *Lark* visited western Newfoundland in 1763, he found a substantial settlement on the old French site at Codroy Island. About a hundred people lived there in homes built directly on the island. This made Codroy a substantial settlement; it was also the only settlement on the west coast, and in contrast to the earlier communities, Codroy in 1763 did not yet include many women; Thompson counted only five families.[74] Nevertheless, most of the inhabitants stayed at Codroy year round, supporting themselves in a familiar pattern of fishing in the summer and trapping beaver, otter, marten and fox during the off-season. Most of the fishing was conducted locally at Codroy, though some of the inhabitants fished at various locations as far north as the Bay of Islands.[75]

---

[71]TNA/PRO, CO 195/9, 330-356, Board of Trade to the King, 20 April 1764.

[72]The Board of Trade must have been quite aware of this fact. *Lark*, Captain Samuel Thompson, had spent two months patrolling the coast between Port aux Basques and Port aux Choix in 1763; his report to his superiors indicated that Codroy had belonged to the French until "the beginning of the late Warr." TNA/PRO, ADM 1/2590, No. 4, Captain Thompson's Letters, Thompson's report to Admiralty Secretary Philip Stephens, 12 March 1764. The most detailed description of the Codroy settlement appears in his "Remarks" in National Maritime Museum, Greenwich (NMM), Graves Papers (GRV) 105, Answers to Heads of Inquiry, 1763. Unless otherwise indicated, the descriptive details concerning the Codroy settlement after 1760 all come from Thompson's "Remarks" and "Answers."

[73]TNA/PRO, CO 195/9, 286-287, Board of Trade, draft of instructions for Hugh Palliser, No. 12, 10 April 1764.

[74]NMM, GRV/105, No. 46, "Answers to Heads of Inquiry, 1763," especially "Remarks" and "Answers" of Captain Thompson.

[75]In 1764, Governor Palliser would learn that some unidentified Englishmen had overwintered in the Bay of Islands in 1762; see his remarks in British Library,

A significant number of the people were Irish, and many were former residents of St. Pierre until displaced by the British decision to return that place to France in 1763; others were from various parts of Newfoundland. At least twenty-six servants, however, were brought from Louisbourg and Halifax – it would be nice to think that some servants from Louisbourg had names like Sabot, Bourny and Commaire, but thus far no records of names have been found. The willingness of metropolitan merchants to maintain a permanent fishing station through the employment of imported and residential labour had been essential for Codroy's survival in its earlier incarnations; this was also the case in 1763. The settlement's "principal inhabitant" was Jonathan Broom; during the 1750s, he had been linked with Bonavista on Newfoundland's east coast, but by the 1760s Broom was maintaining Codroy as a fishing station, like the Chenus and other predecessors. His employees could also be found at Bay St. George, the Bay of Islands and possibly points beyond that.

Though there was clearly some commerce between St. Pierre and Codroy, Broom supplied most of the settlers' needs; he shipped in salt, clothing and some bread and flour from England, pork and butter from Ireland, while most of the provisions, including pork, bread, flour, rum and molasses, as well as tobacco, salt and cordage, came from New England. Apart from the fishery, the attraction of this part of Newfoundland for Broom and the inhabitants, as it had been for their French predecessors, lay in its abundant resources. Mention has already been made of fur trapping, and visitors also remarked on the wealth of the salmon fishery and the quality of the timber. Captain Thompson observed that "in general the Coast is cover'd with large Trees fit for Ship Building, such as ye various Kinds of Birch, Juniper, Pines & Spruce fitt for masts;" in 1767, at least two vessels were built at Codroy.[76]

In 1783, the entire region fell under French control, following the redefinition of the French Shore. The eastern limits of the French Shore were moved west to Cape St. John, and a similar adjustment was made to its western limits, which now extended all the way to Cape Ray. Though their claims in 1763 to the west coast had not been pursued vigorously, the French had worked long and hard to settle the question of whether the fishery on the Treaty Shore was exclusively French or whether it had to be shared with the British – the Treaty of Utrecht had not made this clear. The French had failed to prevent the penetration of Notre Dame Bay during the 1720s and 1730s by British fishermen and settlers and did not insist that their fishing privileges on the Treaty Shore were exclusive, rather than concurrent, until the 1760s. Since

---

Additional Manuscripts (BL Add.) 17,693a, "A Plan of the Bay of Three Islands in Newfoundland...taken on board HMS *Guernsey*," June 1764. Presumably these overwinterers were connected with the settlement at Codroy.

[76]NMM, GRV/105, Thompson, "Remarks;" and Head, *Eighteenth Century Newfoundland*, 165.

it was too late by then to remove the British from Notre Dame Bay, France chose to abandon it and to re-define the Treaty Shore to include the relatively empty west coast. In this way, the French hoped to acquire a second chance to develop their Treaty Shore fishery exclusively.[77]

As a result of this re-definition of the Treaty Shore and of government subsidies and bounties, the French fishing industry was quickly re-established on the west coast. But this meant discouraging any English settlers found here to forestall a repetition of what had happened in Notre Dame Bay. The settlements on the west coast therefore found themselves under pressure from the French to leave. In 1785, the French destroyed two salmon posts in Bay St. George and tried to force settlers in the area to leave. Broom (or perhaps by then his son) re-located to the Burin Peninsula, though it is not clear what, if anything, happened to the people living at Codroy. French hostility to settlement in western Newfoundland was interrupted by the French Revolutionary and Napoleonic wars after 1793, with the result that during the next twenty years the commercial and demographic growth in the region continued.[78] By the time the French returned after 1815, it was too late to reverse this growth. Thus, the single greatest contrast between the earlier French incarnations of Codroy and its eventual English one was in the effect that war had on the survival of settlement. Conditions of war had always forced the French inhabitants to abandon their homes, whereas wartime conditions allowed settlement to persist once Codroy's identity became English. Today, the French settlements at Codroy in the early eighteenth century are largely forgotten, except for the many French and Basque place names that can still be found on the map.

---

[77]Brière, "Pêche et politique," 175-183; and James K. Hiller, "The Newfoundland Fisheries Issue in Anglo-French Treaties, 1761-1783" (Unpublished paper presented at the Ninth Atlantic Canada Studies Conference, St. John's, Newfoundland, May 1992), 10-11.

[78]The settlement history of western Newfoundland after 1783 is discussed in Mannion, "Settlers and Traders," in Mannion (ed.), *Peopling of Newfoundland*, 234-279.

The Illicit Trade in English Cod into Spain, 1739-1748[1]

In April 1740, the merchants of Bayonne complained to the Chamber of Commerce that a French vessel had arrived with a quantity of dry English cod. It had attempted without success to deliver this fish at the Spanish ports of Bilbao and San Sebastián before making for Saint-Jean-de-Luz, where it discharged part of its cargo. Now it had arrived at Bayonne, where it would remain briefly before renewing its efforts to carry the fish to Spain. Because Spain and England were at war, the normally substantial trade in English cod between those countries had been interrupted. French merchants, whose own cod trade with Spain had diminished considerably since the end of the previous century, had welcomed the war as an opportunity to regain their position.[2] To discover that English cod continued to penetrate the Spanish market was quite alarming; to learn that French dealers were responsible was a scandal. The Bayonnais merchants therefore demanded that appropriate steps be taken to put a stop to the commerce. But while the authorities were sympathetic to the complaint, effective countermeasures were difficult to implement. One year later the *contrôleur-général des finances* learned that French participation in the traffic had persisted; he had received a report of a Malouin vessel being lost off Cape Finisterre while making for Cádiz with 5000 quintals of English cod.[3] Although it was apparent by then that neutral traders, including Swedes and Danes but especially the Dutch, were also involved, the bulk of the carriers were French – twelve Malouin vessels were involved in 1741, according to

[1]This essay appeared originally in the *International Journal of Maritime History*, VIII, No. 1 (1996), 1-22. This paper is based in part on research funded by the Social Sciences and Humanities Research Council of Canada, the assistance of which is gratefully acknowledged. An earlier version of this paper was presented at the Fifteenth Meeting of the French Colonial Historical Society in Martinique and Guadeloupe in May 1989 under the title "'Bretons...sans scrupule:' The Family Chenu of St. Malo and the Illicit Trade in Cod During the Middle of the 18th Century." I wish to thank participants at that conference for helpful comments and suggestions.

[2]Archives de la chambre de commerce de Bayonne (ACCB), I.2, No. 14, Bayonne merchants to Chamber of Commerce, 20 April 1740. See also ACCB, I.2, Nos. 12, 13 and 15, President and Directors of the Chamber of Commerce to M. Orry, *contrôleur-général des finances* and to Minister of Marine Maurepas, 20 April 1740, together with Maurepas' reply of 14 May 1740.

[3]ACCB, I.2, No. 17, anonymous memorial, 1741.

one report, while another claimed that French vessels carried as much as 80,000 quintals of English cod annually to Bilbao alone. It was further alleged that some of those vessels carried not only cod but also English cloth.[4] Such reports caused the authorities to redouble their efforts to stop the trade, but success remained elusive.

Several factors account for this failure, not least that Spain seemed to prefer English cod over that produced by France. According to one memorial, this was because English fishermen used Spanish salt, which was judged superior to French salt.[5] Yet of at least equal importance was the determination by some French merchants to make the illicit trade in English cod the latest of several adaptive strategies not just to cope but to thrive in the increasingly difficult French fishery and trade in North America. Those merchants tended not to be Bayonnais or French Basque, whose role in the North American fisheries was by the late 1730s in a seemingly irreversible decline.[6] Rather, as one official explained, "*les Bretons sont les seuls qui S'y Livrent sans scrupule.*"[7] In particular, certain merchants of Saint-Malo appear to have been most active in carrying English cod to Spain.

Saint-Malo's association with the North American cod fishery and trade was as old as the fishery.[8] Bretons had been among the first Europeans to

---

[4]Archives maritimes au port de Brest (AM Brest), 1P[1]/3, No. 80, anonymous memorial, 17 September 1741; No. 86, anonymous memorial, n.d.; and ACCB, I.2, No. 17, anonymous memorial, 1741.

[5]AM Brest, 1P[1]/3, No. 80, anonymous memorial, 17 September 1741. The French were forbidden by law to obtain or use Spanish salt. It should perhaps be noted that Malouin merchants had long pleaded with the authorities for permission to do so; see, for instance, Paris, Archives de la Marine (AM Paris), B[3]/361, Orry to Maurepas, 26 January and 16 March 1733; and B[3]/388, Beauvais le Fer (*armateur* of Saint-Malo) to Maurepas, 3 December 1738. It is therefore possible that the claim that Spain still preferred English cod after 1739 because it had been cured with Spanish salt was in fact the latest ploy to have such salt admitted into the French fisheries.

[6]The best summary of the history of Bayonne, Saint-Jean-de-Luz and Ciboure in the North American cod fisheries and trade is Laurier Turgeon, "La crise de l'armement morutier basco-bayonnais dans la première moitié du XVIIIe siècle," *Bulletin de La Société des sciences lettres et arts de Bayonne*, nouvelle série, no. 139 (1983), 75-91. See also Turgeon, "Les échanges franco-canadiens: Bayonne, les ports basques, et Louisbourg, Île Royale (1713-1758)" (Unpublished mémoire de maîtrise, Université de Pau, 1977).

[7]ACCB, I.2, No. 16, Serilly (Pau) to Bayonne Chamber of Commerce, 6 June 1741.

[8]The touchstone for all studies to date on the French Newfoundland fishery has been Charles de la Morandière, *Histoire de la Pêche Française de la Morue dans*

appear in Newfoundland following the discovery of the immensely productive fishing grounds by John Cabot in 1497.[9] Though several nationalities soon joined them, the French were by far the most numerous by the closing decades of the sixteenth century.[10] And by the time the French fishery peaked late in the seventeenth century, Saint-Malo was its leading participant; during the 1680s, an average of eighty Malouin vessels per year were sent to the fisheries to specialize in the production of the lightly-salted "dry" fish in such demand in Spain, Portugal and the Mediterranean. Throughout the eighteenth century,

---

l'Amerique Septentrionale des Origines à 1789 (3 vols., Paris, 1962-1967). This descriptive study has been succeeded (though not entirely supplanted) by analytical works such as Laurier Turgeon, "Le temps des pêches lointaines: permanences et transformations (vers 1500-vers 1850)," in Michel Mollat (ed.), *Histoire des Pêches Maritimes en France* (Toulouse, 1987), 133-181; and Jean-François Brière, *La pêche française en Amérique du Nord au XVIIIe siècle* (Saint-Laurent, PQ, 1990). Both Turgeon and Brière pay close attention to the emergence of Saint-Malo as the dominant force in the French fishery, but for a general history of that historic port, including its participation in the fishery, see André Lespagnol (ed.), *Histoire de Saint-Malo et du pays malouin* (Toulouse, 1984).

[9]Turgeon, "Le temps," 136. See also Gustave Lanctot, "Thomas Aubert" and "Jean Denys" in George Brown and Marcel Trudel (eds.), *Dictionary of Canadian Biography, Vol. I: 1000-1700* (Toronto, 1966). While a case can be made for a pre-Cabot discovery of Newfoundland by Bristol interests, the argument remains speculative in the absence of convincing evidence. See Alwyn A. Ruddock, "John Day of Bristol and the English Voyages across the Atlantic before 1497," *Geographical Journal*, CXXXII, No. 2 (1966), 225-233; David B. Quinn, "The Argument for the English Discovery of America between 1480 and 1494," *ibid.*, CXXVII, No. 3 (1961), 277-285; and Patrick McGrath, "Bristol and America 1480-1631," in K.R. Andrews, N.P. Canny and P.E.H. Hair (eds.), *The Westward Enterprise: English Activities in Ireland, the Atlantic and America, 1480-1650* (Liverpool, 1978), 81-102.

[10]In 1578, Anthony Parkhurst estimated that there were 150 French, 100 Spanish, fifty Portuguese, thirty to fifty English and twenty to thirty Basque vessels at Newfoundland. Laurier Turgeon argues convincingly that these figures are much too low for the French fleet and, by extension, for the others. In certain years, the notarial records of Bordeaux, La Rochelle and Rouen describe a larger fleet for those ports alone than Parkhurst attributed to all of France. Although the dispersed nature of the fishery, the constant movement of the vessels and the instability caused by war make a conclusive judgment difficult, Turgeon believes that a more accurate measure of the French fleet would be about 500 vessels, which conforms to the estimate of a less-frequently cited observer, Robert Hitchcock. See Turgeon, "Le temps," 137-138; and Turgeon, "Pour redécouvrir notre 16e siècle: les pêches à Terre-Neuve d'après les archives notariales de Bordeaux," *Revue d'histoire de l'amérique française*, XXXIX, No. 4 (1986), 523-550. Note, too, that La Morandière, *Histoire*, I, 220, cites a source which claims that the fleet from the Portuguese town of Aveiro alone numbered 150 vessels in 1550.

one-third of French vessels making for the North American fisheries were Malouin.[11] It was partly because of their experience that Malouin sailors acquired a reputation in the French navy as first-class seamen, and the fisheries acquired a reputation as a "nursery for seamen."[12] It was equally true, as one historian has recently observed, that the fisheries had helped make Saint-Malo a "nursery for merchants," providing both experience and profits to Malouin outfitters, merchants and investors with which to shift into other activities.[13]

If the seventeenth-century Newfoundland fisheries belonged to France generally and to Saint-Malo in particular, those of the eighteenth century belonged increasingly to England. A number of factors brought the French fisheries to a low ebb by 1713 from which they recovered only with difficulty – and with far fewer participants: the protracted period of war between 1689 and 1713 placed heavy demands on the labour supply; the Treaty of Utrecht forced France to give up its claims to, and occupation of, Newfoundland; a failure of the inshore fisheries occurred for more than a decade after 1713.[14] Outfitting at Saint-Malo for the fisheries fell by seventy-five percent between 1689 and 1697. Despite some recovery by the 1730s, the level of Malouin outfitting stood at only half what it had been fifty years before.[15] By the 1780s, Saint-Malo was once again sending vessels to the fishing grounds in numbers matching those of the previous century. Granville began to expand its role in the late 1730s. Yet at the same time the Basco-Bayonnais fishery went into a steady decline, so that by the beginning of the second half of the eighteenth century it

---

[11]Harold Innis, *The Cod Fisheries; The History of an International Economy* (Toronto, 1940; rev. ed.; Toronto, 1954), 127; and Jean-François Brière, "Saint-Malo and the Newfoundland Fisheries in the 18th Century," *Acadiensis*, XVII, No. 2 (1988), 132-133.

[12]Jean-François Brière, "Pêche et politique à Terre-Neuve au XVIIIe siècle; la France véritable gagnante du traité d'Utrecht?," *Canadian Historical Review*, LXIV, No. 2 (1983), 168-170. See also John S. Bromley, "The Trade and Privateering of Saint-Malo during the War of the Spanish Succession," *Transactions of the Société Guernésiase*, XVII (1964), 631-647.

[13]Lespagnol (ed.), *Histoire*, 136; and Bromley, "Trade and Privateering," 283-284.

[14]By the mid-eighteenth century, according to one recent estimate, of more than 1.5 million quintals of cod caught in the North American fisheries, England accounted for 600,000, New England for 400,000 and France for only 350,000; Christopher Moore, "The Markets for Canadian Cod in the Eighteenth Century: France's Cod Trade and the Problem of Demand" (Unpublished paper presented at the Annual Meeting of the Canadian Historical Association, University of Guelph, 1984), 3.

[15]Lespagnol (ed.), *Histoire*, 113 and 132-134; and Brière, "Saint-Malo," 132.

was a spent force.[16] The growth and vigour of Saint-Malo and Granville had thus been gained, at least in part, at the expense of others, and the resurgence of the French fisheries signified not so much a recovery as a transformation, as Saint-Malo and Granville engrossed what other French participants in the Newfoundland fishery abandoned.[17]

Moreover, much of the renewed activity was linked to a rapidly expanding bank fishery after mid-century that produced a "wet" cure suited only for the domestic market rather than traditional markets in southern Europe. Only the inshore boat fishery produced the preferred "dry" cure. While it, too, showed significant recovery, Malouin merchants were increasingly uncompetitive with their British rivals in the Iberian and Mediterranean markets. Several factors were responsible, not least France's steady loss of territory in North America which caused a proportionate reduction in the French resident fishery, thus denying the merchants an alternative source of fish when war disrupted the migratory fishery. It also diminished the extent of coast available for shore facilities, without which there could be no "dry" cure. French merchants were forced to abandon Iberia in favour of Marseille, which serviced the protected domestic market in southern France; very little fish was re-exported from there to other Mediterranean ports.[18]

In Saint-Malo, the crisis in the industry was but one facet of a more general predicament in overseas commerce. According to Andre Lespagnol, the South Sea and Indian Ocean trades, as well as wartime privateering, all ended between 1713 and 1720. By the mid-1730s, traffic in the major trades had diminished by one-third since the 1680s. A number of prominent mer-

---

[16]See particularly Turgeon, "La crise," 78.

[17]By the 1780s Saint-Malo and Granville formed what Brière called a "cod-fishing oligopoly" controlling seventy percent of the French fleet in North America; Jean-François Brière, "The Ports of St. Malo and Granville and the North American Fisheries in the 18th Century," in Clark G. Reynolds (ed.), *Global Crossroads and the American Seas* (Missoula, MT, 1988), 17. See also Brière, "The French Codfishing Industry in North America and the Crisis of the Pre-Revolutionary Years, 1783-1792," in Patricia Galloway and Philip P. Boucher (eds.), *Proceedings of the Fifteenth Meeting of the French Historical Society, Maritinique and Guadeloupe, May 1989* (Lanham, MD, 1992), 201-210; and Brière, "Le trafic terre-neuvier malouin dans la première moitié du XVIIIe siècle 1713-1755," *Histoire sociale/Social History*, XI, No. 2 (1978), 362.

[18]Brière, "Le trafic," 362-363; Brière, "Saint-Malo," 134-135; and Moore, "Markets," 12-15. The disappearance of the Basco-Bayonnais fishery as a significant factor after 1750 meant that by the second half of the eighteenth century Bilbao, once a major market for the French Basque fishery, no longer received any French fish. See Roman Basurto Larraiiaga, *Comercio y Burguesia Mercantil de Bilbao en la Segunda Mitad del Siglo XVIII* (Bilbao, 1983), 231.

chants used their wealth to shift out of commerce altogether; others who remained commercially active avoided the fishery.[19] While Saint-Malo did not suffer the fate of ports like Dieppe, Honfleur or Les Sables d'Olonne, which withdrew from overseas cod fishing, it was unable to follow ports like Nantes, Le Havre or Bordeaux into more profitable transatlantic trades.[20] In short, as Brière concludes, while Saint-Malo's fishing industry would recover, Malouins who remained committed to it "did so reluctantly, as if they had no choice."[21]

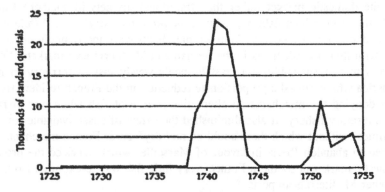

Figure 1:   Re-Exports of Cod from Marseille to Spain, 1726-1755

Source: C. Grant Head, Christopher Moore and Michael Barkham, "The Fishery in Atlantic Commerce," in R.C. Harris (ed.), *Historical Atlas of Canada, Vol. I: From the Beginning to 1800* (Toronto, 1988), plate 28. Reprinted with the kind permission of the University of Toronto Press.

---

[19]Lespagnol (ed.), *Histoire*, 153-154; and Brière, "Saint-Malo," 132. In examining the wealthy and powerful family Picot, Lespagnol noted that it increasingly shifted away from direct involvement in outfitting and shipping and into finance. Its disinterest in the Newfoundland fish trade "*confirme que Terre-neuve, en pleine crise après la Paix d'Utrecht, n'est plus un secteur porteur, un terrain d'accummulation pour le grand négoce malouin.*;" André Lespagnol, "Une dynastie marchande malouine: les Picot de Clos-Rivière," Société d'Histoire et d'Archéologie de l'Arrondissement de Saint-Malo, *Annales 1985*, 234-235. Both Brière and Lespagnol note how capital accumulated in overseas commerce was reinvested in venal offices, real estate and marital alliances with ancient Breton nobility by wealthy families anxious to reduce or even escape their bourgeois origins.

[20]Jean-François Brière, "The Port of Granville and the North American Fisheries in the 18th Century," *Acadiensis*, XIV, No. 2 (1985), 105. Lespagnol attributes Saint-Malo's failure to follow other ports into Caribbean trade not to a lack of exports but to the lack of a significant local market for Caribbean imports; Lespagnol (ed.), *Histoire*, 134.

[21]Brière, "Saint-Malo," 138.

It is in this context of uncertainty and crisis that we must examine Malouin participation in the trade in English cod to Spain at the beginning of the Anglo-Spanish War. As is usually the case when examining a clandestine trade, there are insufficient data to describe with confidence its precise volume or extent. Nevertheless, there are sufficient traces in the documentary record to suggest some broad generalities and enough indicators not only to identify at least one Malouin family involved in the trade but also to offer some tentative conclusions concerning their activities.

Spain, the target for this clandestine trade, was the major European consumer of "dry" cod. Indeed, as the eighteenth century progressed, Iberia became dependent upon England to satisfy its demand for dry cod; England exported about 400,000 quintals to Spain and Portugal in 1735 alone. About twenty percent of that fish entered Spain through Bilbao, even though it was quite close to the French fishing ports of Saint-Jean-de-Luz and Bayonne.[22] Although Spain had been the principal foreign outlet for French dry cod, France could only supply Spain with about one-tenth of the English volume. As a destination for French dry cod, Spain by the 1730s had become secondary to the domestic market.[23] Then, in 1739 war broke out between Spain and England, causing trade to be interrupted. The price of dry cod fell by twenty-four percent in New England while it rose nineteen percent in Spain.[24] For French merchants, it was a heaven-sent opportunity, not only to make some unexpected profits but also to regain control of a trade which had slipped, seemingly irreversibly, into English hands. Re-exports of dry cod from Marseille to Mediterranean ports which had been serviced by English traders before 1739 soared to more than 50,000 quintals, with half going to Spain.[25] The number of vessels outfitted in Saint-Malo for the "dry" fishery at Newfoundland nearly doubled between 1740 and 1743.[26] Nevertheless, insatiable Spanish demand

---

[22]James Lydon, "Fish for Gold: Massachusetts Fish Trade with Iberia, 1700-1773," *New England Quarterly*, LIV, No. 4 (1981), 543; and Moore, "Markets," 11.

[23]Moore, "Markets," 6-8 and 10-11. The merchants of Bayonne insisted that "*la moruë de la pesche Française n'a d'autre debouché que la Navarre et L'Aragon en Espagne;*" ACCB, I.2, No. 14, Bayonne merchants to Bayonne Chamber of Commerce, 20 April 1740.

[24]Daniel Vickers, "'A Knowen and Staple Commoditie:' Codfish Prices in Essex County, Massachusetts, 1640-1775," *Essex Institute Historical Collections*, CXXIV, No. 3 (1988), 192.

[25]See C. Grant Head, Christopher Moore and Michael Barkham, "The Fishery in Atlantic Commerce," in R.C. Harris (ed.), *Historical Atlas of Canada, Vol. I: From the Beginning to 1800* (Toronto, 1988), plate 28; and Moore, "Markets," 25.

[26]Brière, "Le trafic," 364.

kept well ahead of rejuvenated French efforts to meet it. Consequently, until 1744, when France joined the war, the opportunity to acquire English cod and sell it to the Spanish could not be resisted.

French merchants acquired cargoes of English cod for transport to Spain from at least two sources. One was the English Channel Islands, especially Jersey, with which Saint-Malo and the neighbouring Breton and Norman ports had a long-standing relationship based on legitimate trade in a variety of commodities. During the late seventeenth century, Saint-Malo's trade with the Channel Islands was its second-largest peacetime foreign traffic. This was part of a triangular trading pattern involving the Channel Islands, the Norman and Breton ports and the harbours of southern and southwestern England. Significantly, the English fishery at Newfoundland was based in southwestern England and, to a lesser degree, the Channel Islands, especially Jersey. Equally significant, Bilbao was the preferred destination for Jersey cod. In 1689, the Channel Islands lost the privilege of neutrality on which much of that trade had been based. Legitimate commerce between the islands and France was subsequently disrupted by war, to be replaced by a flourishing smuggling relationship. Channel Islanders were thus quite experienced at trading clandestinely with Saint-Malo and other Breton ports. Finally, for more than a century, English merchants engaged in the Newfoundland fish trade had turned to French ports whenever commercial relations between England and Spain were disrupted.[27] There was therefore a tradition in the Channel Islands of clandestine trade as well as participation in both the Newfoundland fishery and in the fish trade with Spain to encourage an illicit traffic in English salt cod to Spain through Saint-Malo after 1739.[28]

When war erupted that year with England, Jersey could no longer export cod directly to its preferred market in Spain. The price of cod was immediately affected; in Jersey the price of dry cod was half what it was in

---

[27]According to a late sixteenth-century document, Saint-Jean-de-Luz was a receiving port for a variety of English provisions and dry goods "to sarve the newefoundland men...This port sarves when we have a restrainte between Spain and us;" Innis, *Cod Fisheries*, 32, n.11. Pauline Croft indicates that Newfoundland cod was shipped from the West Country to Saint-Malo for transhipment to Spain in the 1580s and 1590s, a development she attributes to the way in which regional priorities and concern for the state of the local economy transcended national concerns; Croft, "Trading With the Enemy 1585-1604," *Historical Journal*, XXXII, No. 2 (1989), 292 and 295-296.

[28]Bromley, "Trade and Privateering," 280-281; Rosemary Ommer, *From Outpost to Outport: A Structural Analysis of the Jersey-Gaspé Cod Fishery, 1767-1886* (Montréal, 1991), 14-15; J.C. Appleby, "Neutrality, Trade and Privateering, 1500-1689," in Alan G. Jamieson (ed.), *A People of the Sea: The Maritime History of the Channel Islands* (London, 1986), 59-60, 74 and 89-90; and Alan G. Jamieson, "The Channel Islands and Smuggling, 1680-1850," in Jamieson (ed.), *People*, 196-199.

Spain.[29] It was therefore perfectly sensible for some Malouin merchants to begin sending their vessels to Jersey, where they would load with cod for re-export to Spain. Usually, the Malouin vessels were partially laden with French cod; presumably this provided them with initial clearances which were used to disguise the rest of the cargo as French. In fact, up to three-quarters of the final cargo would be Jersey-produced cod, although there is also evidence that quantities of cod were moving into Jersey from Poole and other West Country ports to take advantage of the emerging re-export trade to Spain.[30] There is also evidence that buried under the Jersey cod were shipments of English cloth.[31] The Malouin vessels then returned home to secure revised clearances which gave no indication that only a small proportion of the cargo was French. When the vessel subsequently arrived in Spain, the cargoes were presented – and received – as French fish.[32] Some indication of the profits to be made is suggested by the differences in prices. The French cod which first went into the hold cost between thirteen and fourteen *livres* per quintal; the Jersey cod cost the merchant only seven to nine *livres* per quintal. In Spain, the fish fetched a price of fifteen to seventeen *livres* per quintal.[33]

The other source of English fish was directly from North America. There were several ways in which such fish could be acquired. Early in 1744,

---

[29]AM Brest, IP[1]/3, No. 86, anonymous memorial, 1741.

[30]Peter Raban, "Clandestine Trade in the Mid-Eighteenth Century" (Unpublished manuscript, January 1988), 3-4. I wish to thank Canon Raban for sending me this draft which he was preparing for publication in *Transactions of the Société Guernésiase*.

[31]ACCB, I.2, No. 17, anonymous memorial, 1741. "*Il y a aux Isles de Gersey et de Grénésay une quantité prodigieuse de ces morüese que les Maloins font transporter chez Eux et se proposent de faire Embarquer dans le cour de cette année pour L'Espagne. Depuis un an il est entré dans le Seul port de Bilbao pluse de 80 mille quinteaux de morue Angloise que les Vaisseaux françois y ont transportés Et sous ces morües on fait passer beaucoup de Draperie angloise.*"

[32]AM Paris, B[3]/405, 122-124, M. Caze de la Bove (*Intendant du commerce de Normandie et Bretagne*) to Maurepas, 20 August 1741; AM Brest, IP[1]/3, Maurepas to unknown (possibly M. de la Fossingnant, *commissaire-général de la marine* at Saint-Malo), 5 October 1741; Pau, Archives départementales des Pyrénées-Atlantiques (AD P-A), B8730, 41v-42, Louis-Jean-Marie de Bourbon, duke de Penthièvre, *Amiral de France*, to the officiers de l'Amirauté de Bayonne, 30 October 1741. The French authorities identified the Sieur Vallois, owner of the barques *Sainte-Anne*, *Marie-Joseph* and *Saint-Gand*, and the Sieur Poulard, who outfitted the ship *Marquis*, as participants in this trade. The vessel lost in 1741 while making for Cádiz with English cod belonged to the Sieur de la Roche.

[33]AM Brest, IP[1]/3, No. 86, anonymous memorial, n.d.

the Minister of Marine was informed that a Malouin merchant had been sending his ships, not to Île Royale as claimed, but to Placentia, where they would trade with the English.[34] Other reports claimed that French merchants were trading with renegade French fishermen living in small settlements on the southwestern coast of Newfoundland; the merchants exchanged provisions, gear, salt and other necessities for fish and oil. Though the settlements were approximately half French and half Irish, the people they worked for were allegedly English.[35] Finally, Maurepas suspected some Malouin merchants of owning shares in English vessels which were sent to the fishery before making for Spain.[36] The one common thread to these accusations is that they all involved the same Malouin family – the family Chenu.

The family Chenu comprised four brothers: Claude, Sieur Boismory (ca. 1678-17?); Pierre, Sieur Dubourg (ca. 1683-1769); Jacques, Sieur Duchenot (ca. 1687-1758); and Louis, Sieur Duclos (ca. 1694-1774). A fifth Chenu, Jerôme, Sieur Dupré (ca. 1698-17?), may have been a cousin.[37] The family did not belong to the top rank of Saint-Malo's *négociants*; they lacked the wealth, diversity of commercial activity and international associations of the great merchants.[38] Instead, their dealings appear to have been confined to the fishery and related activities – outfitting, shipowning and trade. All appear to have spent their younger adult years as captains of fishing or trading vessels

---

[34]AM Brest, IP¹/3, Maurepas to M. Guillot, *commissaire-général de la marine* at Saint-Malo, 17 April 1744.

[35]Olaf Janzen, "'Une petite Republique' in Southwestern Newfoundland: The Limits of Imperial Authority in a Remote Maritime Environment," in Lewis R. Fischer and Walter Minchinton (eds.), *People of the Northern Seas* (St. John's, 1992), 1-33.

[36]AM Brest, IP¹/3, Maurepas to unknown (possibly de la Fossingnant), 5 October 1741.

[37]Information on the family was compiled from various registers, reports, declarations and other manuscripts housed in the Archives of Fortress Louisbourg National Historic Park (AFL); Archives départementales de l'Ille-et-Vilaine, Rennes (AD I-et-V); Archives de l'arrondissement maritime de Rochefort (AM Rochefort); Archives départementales de la Charente-Maritime, La Rochelle (AD C-M); and archival sources already cited. Also of great use was L'Abbe Paul Paris-Jallobert, *Anciens Registres Paroissiaux de Bretagne (Baptêmes – Mariages – Sepultures): Saint-Malo-de-Phily; Evêché de Saint-Malo – Baronnie de Lohéac – Sénéchaussée de Rennes* (Rennes, 1902) and Paris-Jallobert, *Anciens Registres Paroissiaux de Bretagne (Baptêmes – Mariages – Sepultures): Saint-Malo (Evêché – Seigneurie commune – Sénéchaussée de Dinan)* (Rennes, 1898). The parents of the Chenu brothers were Jean Chenu and Margueritte Porée.

[38]Lespagnol (ed.), *Histoire*, 152-153.

which they either owned themselves or which were owned by a brother. This was fairly typical of all levels of merchants engaged in overseas commerce, although Louis Chenu seemed content to remain a captain-owner throughout much of the 1730s and 1740s, even as his brothers were establishing themselves as merchants of Saint-Malo or its suburb Saint Servan. Claude Chenu appears to have been the most mobile. At various times he was identified as belonging to Granville, Saint-Malo and La Rochelle, although he always maintained both the business and personal sides of his relationship with his brothers. Together, the Chenus maintained fishing stations, owned vessels, sometimes as individuals but usually in partnership with one another, and participated in the truck trade at Île Royale. Their sons provided a large labour pool from which were drawn the captains and junior officers of the Chenu vessels – and the next generation of merchants. This intricate network of shared investment and vessel ownership was fairly characteristic of *"une société familiale,"* by which family connections were used to secure businesses against the many risks of eighteenth-century commerce.[39] In short, the Chenus were fairly typical examples of a family trying to move from ship masters to merchants, shipowners and outfitters, fiercely ambitious and somewhat unscrupulous, determined to emulate those who, through their own success, had demonstrated that it was possible to rise from equally modest means to the highest ranks of Malouin commercial society.[40]

It was presumably because of their ambitions that the Chenus frequently accepted the risks involved in straying across the line between legitimate and illegal commerce. This was most apparent in their involvement with the so-called "renegade" settlements of Cape Ray, Codroy and Port aux Basques on the southwest coast of Newfoundland between 1723 and 1744.[41] These communities were located beyond the official jurisdiction of the French authorities at Louisbourg or the effective control of the English authorities at

---

[39]The importance of family in eighteenth-century French business is discussed in Lespagnol, "Les Picot," 233; and Turgeon, "Les échanges," 86.

[40]Lespagnol (ed.), *Histoire*, 131-132. One such model of success was Noël Danycan, Sieur de l'Epine (1656-1734), the son of a small outfitter who began his career as a captain-outfitter of a Newfoundland fishing vessel. Worth only 15,000 *livres* at the time of his marriage in 1685, he was worth millions twenty years later thanks to investments in the South Sea trade.

[41]The exact location of the Chenu fishing station is uncertain. The names "Port aux Basques," "Cape Ray" and "Codroy" all appear in the documents at various times. These names may have been used interchangeably; the name "Codroy" is itself a corruption of Cape Ray (*Ca' de Ray*). Codroy Island was the most probable location for the Chenu fishing station; small, barren and windswept, its appeal rested on the excellence of the local fishery and the particular qualities of the island that made it ideal for curing fish. See Janzen, "Une petite Republique," 17.

Placentia.[42] Yet they were well within the range of British, Anglo-American and metropolitan French commerce. Consequently, the settlements thrived, their inhabitants attracted by freedom from regulations, fees, official supervision and the law, and the merchants drawn by the opportunity for unsupervised trade with the British. By the mid-1730s the trade was substantial enough to attract the notice of London and Paris, each of which insisted that the settlements were supported by the other's merchants, and neither of which was able to stop it.[43] At the centre of all this activity was the family Chenu.

The Chenu association with post-Utrecht Newfoundland can be traced at least to 1714 when Claude Chenu tried to preserve his fishing station at Fortune at the tip of Newfoundland's Burin Peninsula by swearing an oath of allegiance to the English crown. Yet within a few years, he had abandoned that post in favour of Codroy Island, where he was beyond the scrutiny and interference of both English and French authorities. In 1723 and 1724, he was recruiting resident-fishermen at Scatari, an island off the coast of Île Royale, to work at his fishing post at Codroy.[44] Two such recruits, Basile Bourny and a fellow named Sabot (probably Jean Sabot, who married Basile's sister, Jeanne) had lived on the far side of Fortune Bay from Chenu's fishing post in 1714, and probably knew Chenu from that time; they may even have worked for him then, as they did now at Codroy. The Bournys, Sabots and a few other former residents of Fortune and Hermitage bays, such as the Vincents, Commaires and Durands, provided the nucleus of the Codroy settlement for the next twenty years.[45]

Claude Chenu appears therefore to have played a formative role in establishing the so-called "renegade" fishing settlements of southwestern New-

---

[42]Apart from a visit to Port aux Basques by one of the Newfoundland station ships in 1734 and the expulsion of French fishermen from Codroy and Port aux Basques by British warships from Halifax in 1755, no attempt was made to extend British authority over southwestern Newfoundland before 1763. See Olaf Janzen, "Showing the Flag: Hugh Palliser in Western Newfoundland, 1764," *The Northern Mariner/Le Marin du nord*, III, No. 3 (1993), 4-7 and 8-9.

[43]Great Britain, National Archives (TNA/PRO), Colonial Office (CO) 194/9, 177, "The humble representation of William Taverner...," 2 February 1734; British Library, ADD. Ms. 32,785, ff. 103-104v, Lord Waldegrave's Memorial, 29 May [O. S.] 1734; and TNA/PRO, State Papers (SP) 78/207, 66-67v, J. Burnaby (Paris) to John Courand, 2 March 1735 (N.S.) For a fuller discussion of the settlements and the diplomatic response to their presence, see Janzen, "Une petite Republique," esp. 22-25.

[44]TNA/PRO, CO 194/5/265, census of planters taken by William Taverner during his initial survey of the south coast in 1714; and AFL, AC C¹¹B/7, 71-71v, *ordonnateur* de Mezy (Louisbourg) to Maurepas, 27 November 1724.

[45]Janzen, "*Une petite Republique*," 17-19.

foundland. Undoubtedly, his brothers played a part as well; certainly, in the two decades after 1723, they were instrumental in maintaining them. At first it was principally Claude and Pierre who, as owner-captains, appeared in the records, visiting Scatari Island but maintaining a seasonal fishing station at Codroy; Louis and Jacques played junior roles, often serving as second captains. Very quickly, the seasonal character of the Codroy station gave way to permanent settlement; women made their appearance early in the 1720s and before long, children were being born there. Such growth soon attracted English (or, more probably, New England) vessels to trade with the inhabitants, with the substantial numbers of French fishing vessels on their way into the Gulf of St. Lawrence and up the west coast of Newfoundland, and of course with the Chenus.[46] By then, Jacques had assumed a more prominent role in family activities; indeed, he even came to dominate it. He owned and outfitted many of the vessels; he went frequently to Île Royale; and was the most active Chenu in commerce at Louisbourg. Pierre became increasingly rooted as a shipowner, outfitter, and merchant at Saint Servan, while Claude was based at least in part at La Rochelle, handling affairs there (Chenu vessels often made for the port on their homeward voyage, picking up salt which was always in demand in the fishery). Louis remained a captain-owner and was a frequent visitor at Codroy and Cape Ray by the late 1730s and early 1740s. Presumably he supervised family affairs there.

Chenu commercial activity at Île Royale and southwestern Newfoundland intensified after 1739, when the outbreak of war between England and Spain disrupted the English codfish trade. That disruption coincided with a drastic decline in the productivity of the Île Royale fishery. By 1742, the authorities at Louisbourg claimed that French ships coming either to fish or to trade were obliged to leave with holds only partially filled. Requests to load with English cod, which could be acquired from New England vessels drawn to Louisbourg by its role as an entrepôt, were denied.[47] Yet obviously the directives were ignored; New England fish found some means to re-enter the

---

[46]AFL, AC C11B/12, Governor St. Ovide (Louisbourg) to Maurepas, 14 November 1732; AM Brest, 1P1/2, Maurepas to de la Fossingnant, 31 March 1733. It appears that French vessels heading for the fisheries became accustomed to securing part of their provisions at Île Royale. For those passing through the Cabot Strait, Anglo-American supplies at Cape Ray or Codroy would have been more convenient. The failure of the Canadian harvest in 1743 affected the availability of provisions at Île Royale and enhanced the role of Cape Ray. See AM Brest, 1P1/3, Maurepas to Guillot, 7 February 1744.

[47]AC, C11B/24, Governor Duquesnel and *ordonnateur* Bigot to Maurepas, 17 October 1742. See also Andrew Hill Clark, *Acadia: The Geography of Early Nova Scotia to 1760* (Madison, WI, 1968), 307; and B.A. Balcom, *The Cod Fishery of Isle Royale, 1713-58* (Ottawa, 1984), 15-17 and 50.

Spanish markets because New England cod prices, depressed by the outbreak of the war with Spain, made a substantial recovery between 1741 and 1743.[48] Everything points to the settlements in southwestern Newfoundland as providing one channel; not only was the fishery there reputed to be abundant while that of Île Royale was in a slump, but the Anglo-Americans were long-accustomed to trading with the inhabitants of Codroy. By 1742, the trade in cod between Anglo-American and French traders was reported to be greater there than at Île Royale. By then, however, it was clear that the Anglo-Americans were no longer acquiring cod but instead were supplying it to French buyers.[49]

The likeliest buyers of English cod were Malouin traders. In 1737, *Vandangeur* of Granville, Captain Olivier Grentel, was declared as having made for Niganiche, but was suspected by French authorities of trading at Port aux Basques.[50] *Jason*, owned by La Garande Le Pestour of Granville, departed Saint-Malo in June 1741 with a cargo of salt and provisions, ostensibly to fish at Bay St. George. Its late departure date and the nature of its cargo suggests that trade was its real object, even though its voyage description shows no evidence of such activity.[51] Nevertheless, most vessels which can be identified as appearing at or near southwestern Newfoundland seem to have been on genuine fishing voyages.[52] Only the Chenu brothers had the opportunity to be at the

---

[48]Vickers, "Knowen and Staple Commoditie," 192.

[49]AFL, C[11]B/24, 28-30, Governor Duquesnel and *ordonnateur* Bigot to Maurepas, 17 October 1742; and 111-119, Bigot to Maurepas, 4 October 1742.

[50]Cf. AD I-et-V, 9B/499, and AM Brest, 1P[1]/2.

[51]AD I-et-V, 9B/501, 23 October 1741.

[52]Fishing vessels from Bayonne, and especially Saint-Jean-de-Luz, were the most frequent visitors to this stretch of coast; Turgeon, "La crise," 81-84; and AM Rochefort, sous-série 13P[8], "Rôles des bâtiments de commerce (1723-1926)." There was also a small but steady movement of Malouin vessels on their way to fish at Bay St. George, the Bay of Islands ("*la Baie des Trois Îles*") and Grand Bay (the Straits of Belle Isle). Most had the tonnage and crew size normally associated with the migratory shore fishery. See, for instance, voyage descriptions in AD I-et-V, 9B/171, 419, 500 and 501, for *Alexandre* (150 tons, seventy-three men in 1740); *Mars* (200 tons, 104 men in 1739 and 108 men in 1740); and *Due d'Aumont* (ninety tons, forty-six men in 1740). Yet even *Mars* was suspected of such trade; AM Brest, 1P[1]/2, Maurepas to de la Fossingnant, 4 March 1738. Indeed, while its voyage description for 1740 indicates that *Mars* was at the Bay of Islands continuously between 12 May and 22 August, it had clearly stopped at Port aux Basques long enough to leave its chaplain and possibly others. For a few days beginning 17 May, the chaplain, Louis Colas, performed marriages and baptisms among the inhabitants there; Paris, AN Section Outre-Mer (ANO), G[1]/410, fols. 1-5.

centre of this trade, thanks to their role in the establishment of the community at Codroy and their commerce with Louisbourg. Only Chenu vessels, too small for fishing but perfectly suited for trade, appeared year after year at Codroy and Cape Ray – consistently enough that the authorities occasionally took sufficient notice of their activities to try to stop them.[53] Finally, one family – the Chenus – was consistently suspected by French officials of acting as a front for Anglo- American shipments of cod to Spain.[54]

Attempts to act on those suspicions and to stop Chenu activity invariably failed. This may well have been because French authorities found themselves in a dilemma, in which the temptation of restoring the position of France in the fish trade with Spain outweighed any desire to terminate an illicit activity. In 1740, the French consul at Barcelona reported to the *contrôleur-général* that French commerce there was growing daily ever since England and Spain had gone to war; that before 1739 the English fish trade with Spain measured nearly a million fish annually; and that in the present circumstances, it should be easy for Breton merchants to take that trade back (*"il seroit facile dans les circonstances présentes aux négociants de Bretagne de s'emparer de ce commerce..."*).[55] Similarly, at a time when the fishery at Île Royale was in serious difficulty, the authorities at Louisbourg may have been prepared to turn a blind eye to illicit trade in English cod if the well-being of the local economy was enhanced. It was all a confirmation, perhaps, that regulation and supervision of overseas commerce and possessions were more theoretical ideals than practical measures in the eighteenth century. How else can we explain away the fact that, despite a few instances when they were challenged by the

---

[53]AC, C¹¹B/7, *ordonnateur* de Mezy (Louisbourg) to Maurepas, 27 November 1724; AM Brest, IP¹/2, Maurepas to de la Fossingnant, 31 March 1733; and AM Brest, 1P¹/3, Maurepas to (de la Fossingnant?), 5 October 1741.

[54]In 1741, the French Minister of Marine complained *"J'ay été inforrmé...que les Srs Chenu du Chesno [Jacques Chenu], Dubourg Chenu [Pierre Chenu] son neveu et du hamel Allain prétent leurs noms a des Batimens Anglois dans lesquels ils se disent interessés d'un quart et envoyent a la pesche ces Batimens qui passent ensuite en Espagne."* AM Brest, 1P¹/3, Maurepas to unknown (possibly de la Fossingnant), 5 October 1741. During the 1730s, one of the Chenu vessels engaged in the fish trade was *King George* (see appendix). So unlikely a name for a French vessel suggests that the family Chenu already had commercial connections of some kind with Anglo-American interests, although it must also be conceded that a significant proportion of French ships and vessels in the Newfoundland fishery and trade had been built in New England and purchased through Île Royale. For instance, according to Laurier Turgeon, 17.9 percent of a sample of 318 vessels belonging to merchants of Bayonne and French Basque ports engaged in the Newfoundland fishery and trade were New England-built; Turgeon, "La crise," 86-88, esp. *carte* 2.

[55]Brière, "Le trafic," 364-365.

authorities, there is no evidence that the Chenus ever came into serious difficulties because of their continuous association with fishing settlements which
were, after all, an embarrassment to government? The family tried, of course,
to be discreet about some of their appearances in southwestern Newfoundland.
Thus, the official voyage summary submitted by Louis Chenu for *Légère* in
1741 gave no indication that there had been any ports of call between Saint-
Malo on 3 June and his arrival at Laurembec on Île Royale, an otherwise unaccountable two months later. Yet the marriage and baptismal records of a
Recollet priest who visited Codroy in July reveal that Louis was there, serving
both as a godparent and a witness. Other Chenu vessels made equally lengthy
– and therefore suspicious – ocean crossings, but cannot so easily be proven to
have put in at one of the ports of southwestern Newfoundland.[56] Nevertheless,
the fact remains that the Chenus seem never to have tried very hard to disguise
their activities. Cape Ray appeared in the port records of Saint-Malo as the
intended destination of two of Jacques Chenu's vessels in 1743 and 1744,
while Codroy was the intended goal of another in 1743; in that same year
Claude Chenu was co-owner of a vessel destined for Cape Anguille, which
looms over Codroy Island. Though the Chenus expressed remorse over complaints about such behaviour, and offered assurances that it would not happen
again, there was never any indication that they felt compelled to mend their
ways.

        We are left, then, with a picture of the Chenus as an ambitious, mid-
dle-rank merchant family of Saint-Malo who used every opportunity to advance their personal and family fortunes in the Malouin fishery and fish trade
in the decades following the War of the Spanish Succession. Their ambition
led them to become involved with the establishment and maintenance of illegal
fishing posts on the coast of southwest Newfoundland, and to use them to support commerce with the inhabitants and with Anglo-American traders long
after access to that region had been officially denied to France. When the demand for English cod persisted in Spain following the outbreak of war in
1739, the Chenu brothers adapted their commercial activities in southwestern
Newfoundland to satisfy that demand. Their vessels picked up whatever fish
they could from the inhabitants or from Anglo-American vessels that were now
drawn in greater numbers. The fish may have been carried to Louisbourg to be
disguised with the appropriate documents as French fish, before being carried
to Spain. French authorities were aware that the Chenus were involved in the
traffic in English cod to Spain, but their efforts to interdict it were ineffective,
possibly because it served both imperial and regional interests to tolerate rather

---

        [56]Hirondelle (or *Irondelle*): AD I-et-V, 9B/501, 13 March 1743, and 9B/420,
157, 21 March 1744; *Achille*: AD I-et-V, 9B/171, 7 June 1743; *Saint-Ursin*: AD I-et-
V, 9B/171 and 9B/420, 22 May 1743; and *Légère*: cf. AD I-et-V, 9B/501, 5 December
1741 with ANO, G^1/407, registre 1, fol. 76, "Baptemes fait à lisle de Cadray 1741."
See also the appendix.

than suppress such trade. The illicit traffic in English cod would only end in 1744 when the Spanish government began to allow Dutch vessels to bring it into the country. Spanish markets were soon flooded, which forced prices below the point where French merchants could compete, whatever the source of their fish.[57] 1744 was also the year in which France entered the war against England, thereby exposing its fisheries to attacks by English privateers. Before the end of 1745, Louisbourg had fallen into British hands. There would be no French fishery or fish trade for the next three years.

Much of this essay has of necessity depended upon supposition. It is in the nature of any clandestine trade to leave no obvious traces in official records. The activities of the Chenu brothers must therefore be surmised from mostly circumstantial evidence. The picture which emerges is, however, consistent with what we know about the changing fortunes of the French fisheries in North America, about the commercial history of Île Royale, about the small but persistent settlements in southwestern Newfoundland, and about the channelling of English cod into Spain after 1739 on French vessels. Two points, perhaps, can be suggested by that picture. One is that much remains to be pieced together about the French fisheries and the related trades. True, a rich documentary legacy has permitted historians in recent years to describe with considerable accuracy the patterns of investment, shipping, employment, and trade related to the French fisheries. And in its general dimension, that description is only reinforced, not altered, by the activities of the family Chenu. Yet those activities also suggest that in detail, the volume of fish produced and carried to market may in certain situations and at particular times have been significantly different from what the records seem to tell us. Similarly, the movements of vessels were not always as claimed. The opportunities for clandestine trade in the eighteenth century were numerous, and merit further investigation, not only for what they tell us about commercial patterns but also for what they reveal concerning the limits of metropolitan and colonial authority in an age of "mercantilism."

Second, when historians examine the families who invested in the fisheries, there is an understandable tendency to focus upon the more powerful *négociants*, if only because the records they left are more likely to have survived. And yet, while the investigation of lesser families like the Chenu of Saint-Malo may be much more tedious and difficult, it may also be ultimately just as rewarding. It has been argued that by mid-century some of the wealthier *négociants* of Saint-Malo were seeking ways to distance themselves from commerce generally and the fisheries in particular.[58] Is it not necessary to learn

---

[57]ACCB, I.2, No. 27, M. Sourcade to Bayonne Chamber of Commerce, 9 November 1745.

[58]This argument is articulated in Lespagnol (ed.), *Saint-Malo*, 154. It is personified through his study of "Les Picot," esp. 233-235.

how and by whom the vacuum created by that withdrawal was filled? The Chenu family used the illicit traffic in English cod to Spain as an adaptive strategy to survive the difficulties burdening the French fisheries after 1713 and to climb into a more secure financial and social rank. Though the Chenus suffered considerable financial hardship during the 1744-1748 war, family members were still engaged in the fisheries in the 1780s.[59] Given the trials and tribulations the Malouin fisheries continued to face, their survival in that trade, and the means by which this was possible, merits appreciation.

---

[59]Brière, "Saint-Malo," 137.

Appendix

Vessel (Tonnage/Crew)	Principal Owner	Destination	Date
St. Martin (80/19)	Claude Chenu	Île Royale	04/1726
George (80/16)	Claude Chenu	Scatari	1727
King George (80/15)	Claude Chenu	Louisbourg	05/1739
St. Ursin (90)	Claude Chenu		1741
St. Ursin (90)	Claude Chenu	Cape Anguille	1743
Martin Gallere (70/14)	Pierre Chenu	Île Royale	1726
Martin Gallere (70/?)	Pierre Chenu	Scatari	1727
Quatre Frères (25/6)	Pierre Chenu	Île Royale	05/1734
Heureuse Union (100/19)	Pierre Chenu	Cádiz	11/1734
Margueritte (90/16)	Pierre Chenu	Île Royale	03/1735
Margueritte (80/14)	Pierre Chenu	Île Royale	06/1737
Jacques (70/10)	Pierre Chenu	Gaspé	07/1740
Jacques (70/10)	Pierre Chenu	Gaspé	07/1741
Jacques	Pierre Chenu	Trois Îles	04/1742
Heureux Succes (80/10)	Louis Chenu?	Île Royale	1733
Légère (50/10)	Louis Chenu	La Rochelle	05/1739
Légère (50/9)	Louis Chenu	Île Royale	05/1740
Légère (50/8)	Louis Chenu	Île Royale	06/1741
Légère (50/9)	Louis Chenu	Cape Breton	04/1742
Légère (50/9)	Louis Chenu	Île Royale	1743
Margueritte Anne		Île Royale	07/1728
Heureuse Union (100/22)	Jacques Chenu	Cádiz	06/1734
King George (80/14)	Jacques Chenu	Île Royale	05/1735
Heureuse Union (100/17)	Jacques Chenu	Île Royale	1735
King George (70/16)	Jacques Chenu	Banks	03/1737
Heureuse Union (100/19)	Jacques Chenu	Île Royale	1737
Heureuse Union (150/27)	Jacques Chenu	Île Royale	1740
Hirondelle (50/?)	Jacques Chenu	La Rochelle	10/1740
Hirondelle (50/10)	Jacques Chenu	Île Royale	05/1741
Hirondelle (50/10)	Jacques Chenu	Bordeaux	11/1741
Heureuse Union (150/26)	Jacques Chenu	Cádiz	07/1741
Expedition (70/8)	Paul Liron/ Jacques Chenu	Île Royale	1742
St. Ursin (90/15+1)	Jacques Chenu	La Rochelle/Nfld	03/1742
Hirondelle (50/10)	Jacques Chenu	Île Royale	04/1742
Reine des Anges (80/15+45)	Jacques Chenu	Île Royale	04/1742
Heureuse Union (100/30)	Jacques Chenu	Cádiz	06/1742
Benediction (70/8)	Jacques Chenu	Oporto	06/1742
Margueritte (120/15)	Jacques Chenu	Gaspé	1742
Benediction (70/8)	Jacques Chenu	Bilbao	03/1743
Benediction (70/8)	Jacques Chenu	Oporto	03/1743
Hirondelle (60/11)	Jacques Chenu	Codroy	03/1743
Reine des Anges (80/15)	Jacques Chenu	Île Royale	04/1743
Hirondelle (50/8)	Jacques Chenu	Granville	11/1743

Vessel (Tonnage/Crew)	Principal Owner	Destination	Date
Achille (100/19)	Jacques Chenu	Cape Ray	1743
Hirondelle (?/?)	Jacques Chenu	Cape Ray	03/1744
Georges Bernard (110/18)	Jacques Chenu	Louisbourg	1744

Note: Twenty-five percent of *Expedition* was owned by Jacques Chenu

Sources: See text.

Un Petit Dérangement: The Eviction of French Fishermen from Newfoundland in 1755

One of the more traumatic events in Canadian history – the deportation of the Acadian people from their homes in Nova Scotia – began in the late summer and early autumn of 1755. Something like six or seven thousand Acadian men, women and children were rounded up by Anglo-American and British military and naval personnel, herded onto transports and shipped off to various destinations throughout the North Atlantic while their homes were put to the torch and their livestock slaughtered. It was a remarkable event, described in a recent study of the Seven Years' War in North America as "perhaps the first time in modern history [that] a civilian population was forcibly removed as a security risk."[1] The event certainly remains controversial to this day; although the Acadians had been under British control since 1710 and had resisted pressure to swear an oath of allegiance to the British crown, no one anticipated so drastic a reaction to their stubbornness a full generation later. Moreover, England and France were nominally still at peace in 1755; true, hostilities had already commenced in some parts of North America, but war would not formally be declared until 1756. Historians therefore still search for reasons and meaning to "*le grand dérangement*," while for the modern-day Acadian descendants, it remains both a formative event in the evolution of their culture and proof of English perfidy.[2]

Yet this mid-eighteenth-century act of "ethnic cleansing" was not as unprecedented as one might think. In 1744, as England and France had been about to formalize hostilities in the last war, an attempt had been made to ex-

---

[1]Fred Anderson, *Crucible of War: The Seven Years' War and the Fate of Empire in British North America, 1754-1766* (New York, 2000), 113.

[2]A comprehensive study of the events leading to the Acadian deportation is provided by Geoffrey Plank, *An Unsettled Conquest: The British Campaign against the Peoples of Acadia* (Philadelphia, 2001), while a succinct and useful overview is provided in Stephen E. Patterson, "1744-1763: Colonial Wars and Aboriginal Peoples," in Phillip A. Buckner and John G. Reid (eds.), *The Atlantic Region to Confederation* (Fredericton, NB, 1994), 142-147. An excellent introduction to the history of the Acadian people continues to be Andrew Hill Clark, *Acadia: The Geography of Early Nova Scotia to 1760* (Madison, WI, 1968). A useful analysis of the decision by Nova Scotia's colonial government to remove the Acadians is provided in George A. Rawlyk, *Nova Scotia's Massachusetts: A Study of Massachusetts-Nova Scotia Relations, 1630 to 1784* (Montréal, 1973), 211-213.

pel French inhabitants from Isle Saint Jean (today's Prince Edward Island).[3] And as this paper will show, in August 1755, just a few weeks before the Acadian deportation began in Nova Scotia, a *"petit dérangement"* took place when French fishermen working and living on the coast of southwestern Newfoundland were forcibly removed by English warships. We shall see that one man links all three events, although only additional research will enable us to determine the extent to which his involvement was more than just an historical coincidence. This essay will concentrate on the Newfoundland incident, giving particular attention to how and by whom that expulsion was carried out. I shall argue that the Newfoundland event was driven in part by the way the French and English fisheries were perceived by naval authorities, by the circumstances and opportunity of the moment and possibly by the personal experiences, if not the motives, of the man who directed the operation.

In 1755, as they were edging ever closer to war, both England and France had strong views about the worth of their cod fisheries in North America. As a nursery for seamen as well as a source of national wealth the fishery was so highly prized by both countries that neither would willingly give it up, either in whole or in part.[4] It therefore stood to reason that if one's own fishery were an economic and strategic asset, then a strike against the rival fishery might weaken the capacity of the enemy to engage in war. For instance, in 1706, during the War of the Spanish Succession, three British warships stationed in Newfoundland, together with a contingent of soldiers from the garrison at St. John's, had carried out an operation against the French inshore fishery in the "Petit Nord," as Newfoundland's Northern Peninsula was then known.[5] In 1744, during the War of the Austrian Succession, New England privateers attacked the French bank fishing fleet with considerable success. The privateers were then re-fitted with the assistance of the Royal Navy's Newfoundland station ships and provided with Royal Marines before proceed-

---

[3]Samuel G. Drake, *A Particular History of the Five Years French and Indian War in New England and Parts Adjacent...1744 to...1749, Sometimes called Governor Shirley's War* (Albany, NY, 1870; reprint, Charleston, SC, 2012), 112. I am most grateful to Dr. W.A.B. Douglas for bringing this source to my attention.

[4]During attempts in 1761 to negotiate an end to the Seven Years' War, members of the British and French governments independently ventured the same opinion: that the Newfoundland fishery was more valuable than Canada and Louisiana combined "as a means of wealth and power." Compare the observations of the British Board of Trade, cited in G.S. Graham, "Fisheries and Sea Power," in G.A. Rawlyk (ed.), *Historical Essays on the Atlantic Provinces* (Toronto, 1971), 8, with those of the Duc de Choiseul, cited in Max Savelle, *The Origins of American Diplomacy: The International History of Anglo-America, 1492-1783* (New York, 1967), 475n.

[5]Daniel Woodley Prowse, *A History of Newfoundland from the English Colonial and Foreign Records* (London, 1895; reprint, Belleville, ON, 1972), 246-248.

ing to the Petit Nord where they continued their depredations against the French inshore fishery.[6] An assault of some kind on the French fishery in 1755 as the next war loomed would therefore be consistent with past practice.

Nevertheless, the French in Newfoundland were unprepared when the first blow was struck. For one thing, the attack came in August 1755 at a time when England and France were still technically at peace. Moreover, the attack was directed at French fishermen in southwestern Newfoundland, a corner of the island that had been more or less ignored until then by the authorities of both France and England. The English fishery had always been concentrated primarily on the Avalon Peninsula, the easternmost part of the island of New-foundland. A few fishing merchants and fishermen extended their activities as far west as the island of St. Peter's, Fortune Bay and even Hermitage Bay, but their numbers were too small to justify patrols beyond those points by the Brit-ish warships stationed in Newfoundland every summer to observe and protect the fishery. Only twice between 1713 and 1763 did a Newfoundland station ship venture beyond Hermitage Bay: the *Swift* sloop sailed as far as Ramea and Burgeo in 1716 in support of a cartographic survey of the south coast by Lieu-tenant John Gaudy, while in 1734, HMS *Roebuck*, Captain Crawford, made a brief visit to Port aux Basques and Cape Ray to investigate reports of illegal settlement there by French fishermen.[7] Neither visit led to any departure from the prevailing practice of ignoring Newfoundland west of St. Pierre.

It was in part because of English willingness to ignore the region that French fishing crews persisted in fishing in southwestern Newfoundland. This presence and activity was in violation of the Treaty of Utrecht of 1713 by which France not only acknowledged English sovereignty over the island of Newfoundland but also agreed not to allow its people to settle there. France was permitted to maintain a fishery in Newfoundland, but it was to be a sea-sonal one, based in France: each spring French fishermen would cross the At-lantic to fish on the stretch of coast extending from Cape Bonavista on the northeast coast of Newfoundland as far west as Pointe Riche; in the fall, they were expected to quit Newfoundland and return to France.[8]

---

[6]Drake, *Particular History*, 240-241.

[7]For *Swift*'s activities, see Olaf U. Janzen, "'Of Consequence to the Service:' The Rationale behind Cartographic Surveys in Early Eighteenth-Century Newfound-land," *The Northern Mariner/Le Marin du nord*, XI, No. 1 (2001), 1-10. Crawford's visit in *Roebuck* in 1734 is discussed in Olaf U. Janzen, "'*Une petite Republique*' in Southwestern Newfoundland: The Limits of Imperial Authority in a Remote Maritime Environment," in Lewis R. Fischer and Walter Minchinton (eds.), *People of the North-ern Seas* (St. John's, 1992), 1-33.

[8]James K. Hiller, "Utrecht Revisited: The Origins of French Fishing Rights in Newfoundland Waters," *Newfoundland Studies*, VII, No. 1 (1991), 23-39.

Yet when *Roebuck* visited southwestern Newfoundland in 1734, Captain Crawford found a thriving little community of fisherfolk living at Port aux Basques, Cape Ray and particularly at Codroy, in the shadow of Cape Anguille. These settlements appear to have started out as seasonal fishing stations about ten years after the French officially evacuated Plaisance and St. Pierre and began to develop Île Royale (Cape Breton Island) as a fishing colony.[9] Before long, they were overwintering – French Basque ships were leaving fishing crews behind at Codroy Island during the winters of 1723-1724 and 1724-1725, while residential fishermen from Île Royale were also overwintering there by then. Soon they were joined by wives and families from Île Royale; Codroy Island itself was small, flat and barren, but there was a sheltered anchorage between the island and the shore that the fishermen prized, while the nearby Codroy River, which gave access to the interior, enabled the settlers to supplement the summer fishery with gardening in the valley's favourable micro-climate and winter furring. Over time, they were joined by a number of Irish men and women who may have drifted there from Placentia, where a substantial influx of Irish labour had occurred in the 1720s. What had begun as a fishing station therefore became a permanent and growing community, centred on Codroy but apparently with a few residents also at Port aux Basques and Cape Ray. By 1734, Captain Crawford could report that ten families lived at Codroy – a substantial size for an early eighteenth-century Newfoundland community.[10]

Roebuck had been sent to Codroy in response to reports filtering back to London that the French were settling in Newfoundland, but Captain Crawford was assured by the people he found there that they had all sworn oaths of allegiance to the crown and were loyal British subjects. He therefore took no further action against the inhabitants. In 1736, when reports of French fishing and inhabitancy in southwestern Newfoundland persisted, British officials again directed a Newfoundland station ship to investigate. This time, the officer who received these orders – Captain William Parry, HMS *Torrington* –

---

[9]The Bayonne Admiralty registers include references to fishing crews being left at Codroy during the winters of 1723-1724 and 1724-1725; see Archives départementales des Pyrenées-Atlantiques, Pau, série B (supplement), 8724, registres de l'Amirauté de Bayonne, 1722-1725, entries for 12 January 1724 and 9 August 1725. For the Codroy settlement, see Janzen, "*Une Petite Republique.*'"

[10]Great Britain, National Archives (TNA/PRO), Colonial Office (CO) 104/9, 259-v, Lord Muskerry, "Answers to Heads of Inquiry," 1734, esp. Nos. 60-62. A chaplain on a fishing ship visiting the region in 1740 recorded the names of some forty individuals in the course of performing a number of baptisms and marriages; France, Archives nationales, archives d'outre-mer (ANO), G^1/410, No. 12, Louis Colas, "prestre au monier du navire Le Mars," "Extrait de registre de bapteme et de mariage de la poste du Petit Nord nommé Port au Basques," 18 May 1740.

managed to get no further than Placentia before local officials and merchants convinced him that the inhabitants of Port aux Basques had sworn oaths of allegiance to the English government and never engaged in trade with French ships. Instead, he was assured that they traded only with merchants of Poole, England, who operated through the island of St. Peter's.[11] For the next twenty years, imperial authorities were content to treat the fishing and commerce carried on in southwestern Newfoundland with complete indifference.

Yet had they exercised a little more effort, the naval officers stationed in Newfoundland might have discovered that they were being misled – French merchants were very much involved in the fishery and trade at Port aux Basques and Codroy.[12] Most of the fishing vessels in the region came from Bayonne, Saint-Jean-de-Luz or other French Basque ports.[13] There was also a substantial number of vessels belonging to Saint-Malo which made the west coast either a destination for fishing or for trade as they worked their way up to the "Petit Nord."[14] These activities were allowed to flourish into the 1740s thanks to the failure or unwillingness of British authorities to intervene. The result was a commercial and settlement pattern that contravened the terms of the Treaty of Utrecht. Only with the outbreak of war in 1744 did the inhabitants show any sense of insecurity; if the authorities were unwilling to pay them much attention in peacetime, neither were they willing to expend any effort to protect them in time of war. The threat of attack by Anglo-American privateers caused them to abandon their little settlements for the dubious security of French-controlled Île Royale. They returned, however, as soon as peace

---

[11]TNA/PRO, CO 194/10, 25, Capt. Fitzroy Henry Lee to Board of Trade, 25 September 1736. St. Peter's is more familiar to us as St. Pierre, but the name was anglicized while it was under British control during the period 1714-1763.

[12]See, for instance, Olaf U. Janzen, "The Illicit Trade in English Cod into Spain, 1739-1748," *International Journal of Maritime History*, VIII, No. 1 (1996), 1-22.

[13]See Archives de l'arrondissement maritime de Rochefort (AM Rochefort) sous-série 13P⁸, "Rôles des bâtiments de commerce (1723-1926)." The best study of French Basque involvement in the North American cod fisheries and fish trade is Laurier Turgeon, "La crise de l'armement morutier basco-bayonnais dans la première moitié du XVIIIe siècle," *Bulletin de la Société des sciences lettres et arts de Bayonne*, nouvelle série no. 139 (1983), 75-91, esp. 81-84. See also Turgeon, "Les échanges franco-canadiens: Bayonne, les ports basques, et Louisbourg, Île Royale (1713-1758)" (Unpublished mémoire de maitrise, Université de Pau, 1977).

[14]See, for instance, voyage descriptions in Archives départementales de l'Ille-et-Vilaine, Rennes (AD I-et-V), 9B/171, 419, 500 and 501, for *Alexandre* (150 tons, seventy-three men in 1740), *Mars* (200 tons, 104 men in 1739, 108 men in 1740) and *Duc d'Aumont* (ninety tons, forty-six men in 1740).

was restored in 1748. The settlements were soon re-occupied, and their inhabitants began once again to fish and trade with fishing ships from the Basque ports of southwestern France.

And so it was that in the summer of 1755, the French fishing vessel *Deux Maries* of Ciboure was wrapping up a successful summer at Codroy. Since their arrival in early April, the thirty-five-man crew, together with local inhabitants, had produced a mixed cargo of salt cod, green cod, fish oil, cod tongues and other fish products worth 46,550 *livres*.[15] Not far away, *Ste. Catherine* of Saint-Jean-de-Luz was enjoying an even more successful season, having made 2300 quintals of salt cod and a substantial amount of green cod, as well as acquiring a considerable quantity of additional fish through trade with the local inhabitants.[16] Further up the coast at Bay St. George, the crew of the snow *Saint-Jean*, Capt. Martin Hiriboure, was putting the finishing touches on a season that had yielded nearly 40,000 *livres* worth of salt cod and other fish products.[17]

Suddenly, on 21 August, just as the crews of *Deux Maries* and *Ste. Catherine* began loading their cargoes for the return voyage to Europe, an English frigate appeared and began seizing the French vessels and destroying the French shore facilities. This was the twenty-four-gun *Success*, Captain John Rous, one of several warships that had been stationed in Nova Scotia under the command of Captain Richard Spry until May, when Vice-Admiral Edward Boscawen arrived from England with a squadron of warships and instructions to prevent the French from delivering troops or warships to Louisbourg or Québec.[18] Upon learning of the French communities and fishing activities in southwestern Newfoundland, Boscawen sent *Success*, the twenty-

---

[15]Archives de la chambre de commerce de Bayonne (ACCB), I.2, No. 44, declaration of Jean Lafreche, captain of *Deux Maries* of Ciboure, 11 October 1755; and I.2, No. 40, statement of Jean Lafreche, 9 October 1755.

[16]ACCB, I.2, No. 43, declaration of Captain Sansin Halsouet, 10 October 1755; and I.2, No. 47, statement of Capt. Halsouet, 3 September 1755.

[17]ACCB, I.2, No. 46, statement of Captain Martin Hiriboure, 7 September 1755.

[18]Julian S. Corbett, *England in the Seven Years' War: A Study in Combined Strategy* (2 vols., London, 1907; 2nd ed., London, 1918; reprint, New York, 1973), I, 43-45; and Gerald S. Graham, *Empire of the North Atlantic: The Maritime Struggle for North America* (Toronto, 1950), 158-159. For the French perspective of these opening moves in the war at sea, see Jonathan R. Dull, *The French Navy and the Seven Years' War* (Lincoln, NB, 2005), 25-26 and 31.

four-gun frigate *Arundel*, Captain Thomas Hankerson, and the fourteen-gun snow *Vulture* to search out and destroy the French installations there.[19]

John Rous was an excellent choice to command the little expedition. A New Englander by birth, he figured prominently in some of the more note-worthy events in Nova Scotia and Newfoundland history in the 1740s and 1750s. During the previous war he had commanded one of the New England privateers that cruised in Newfoundland waters against the French, first on the Grand Banks and then against the shore fishery on the Petit Nord. Rous had also been involved in the attempt to drive French settlers out of Isle Saint Jean in 1744. In 1745, he had distinguished himself in the operation that led to the capture of Louisbourg and was rewarded with a captain's commission in the Royal Navy – an impressive achievement for a colonial. In 1748, with the 24-gun *Shirley* under his command, Rous escorted two transports and 300 men into the Minas Basin at the head of the Bay of Fundy and occupied the community of Grand Pré in a show of force designed to discourage Nova Scotia's Acadian population from supporting French war efforts. In 1749, he convoyed the settlers who had been sent from England to Nova Scotia to found the new imperial stronghold at Halifax.[20] He subsequently served as senior naval officer on the Nova Scotia station, where he continued to assert British sovereignty in areas where the ambiguities of the boundary between British and French territory led to friction.

Now, as the fragile peace which had been established in 1748 by the diplomats at Aix-la-Chapelle began to disintegrate, Rous would play a prominent role in several of the events that eventually led to the renewal of war in 1756. For instance, early in 1755, in company with several frigates and a sloop, he escorted nearly three dozen transports and more than 2000 men on an expedition to capture Fort Beauséjour, a French fortification which controlled the narrow isthmus between Nova Scotia and the North American mainland. In July, he led an expeditionary force against a French outpost on the site of present-day Saint John, New Brunswick; the French destroyed their

---

[19]TNA/PRO, Admiralty (ADM), 1/481, despatches of Admiral Edward Boscawen, North America, 1755-1760, esp. 67-67v, Boscawen (*Torbay* at St. Helen's) to the Admiralty, 15 November 1755.

[20]W.A.B. Douglas, "John Rous," in George W. Brown, David M. Hayne and Francess G. Halpenny (gen. eds.), *Dictionary of Canadian Biography, Vol. III: 1741 to 1770* (Toronto, 1974), 572-574 (*DCB*); Drake, *Particular History*, 240-241; and John Grenier, *The Far Reaches of Empire: War in Nova Scotia, 1710-1760* (Norman, OK, 2008), 136. I am much indebted to W.A.B. Douglas for providing me with an early draft of his essay on John Rous which has since been published in the *New Dictionary of National Biography*; this essay includes details not found in the earlier *DCB* entry.

defences and fled before Rous and his ships could reach them.[21] Later that month, he participated in the meetings at Halifax where the decision was made to force the Acadians to leave Nova Scotia. Whether it was because of his frequent experiences in supporting operations against the French or his earlier experience in attacking the French fishery in the more remote parts of Newfoundland, it was at this point that Rous was ordered by Boscawen to clear the French out of southwestern Newfoundland.

The three ships in Rous' command departed Halifax on 6 August and set course first for Louisbourg where they briefly joined a squadron under the command of Rear-Admiral Francis Holbourne which had arrived to support Boscawen's operations. Rous then made for the island of St. Peter's, off the south coast of Newfoundland, before his ships proceeded to their respective destinations. *Success* headed directly for Codroy, long the centre for settlement, trade and fishing in southwestern Newfoundland. *Arundel* spent two or three days at Port aux Basques, then joined *Success* at Codroy. *Vulture* cruised the coast east of Port aux Basques before doubling back and heading up the coast to Bay St. George, eventually rejoining the others at Codroy.[22] During the several days spent in these activities, the three warships seized a number of French vessels, including *Deux Maries*, *Ste. Catherine* and *Saint-Jean*, together with all the fish and fish products they had made or bought that season. They also put the shore facilities, fishing craft and habitations in five harbours to the torch.[23] One small ship, *Notre Dame du Rozaire*, managed to escape, but only by abandoning half its fish and all the gear belonging to its crew.[24] The crews of the captured fishing ships were eventually allowed to make their way up the coast in whatever conveyances they could find until they reached Port

---

[21]Geoffrey Plank, "New England Soldiers in the St. John River Valley, 1758-1760," in Stephen J. Hornsby and John G. Reid (eds.), *New England and the Maritime Provinces: Connections and Comparisons* (Montréal, 2005), 60, based on Council minutes, 15 July 1755, in Thomas B. Akens (ed.), *Selections from the Public Documents of the Province of Nova Scotia* (Halifax, 1869), 258-259.

[22]TNA/PRO, ADM 51/940, II, *Success* log, Capt. Rous; ADM 51/3770, iii, *Arundel* log, Capt. Thomas Hankerson; and ADM 52/1092, iii, *Vulture* master's log, Mr. Joseph Marriott.

[23]In addition to the three French Basque ships already identified, the English seized a snow of Saint-Malo and at least two ships bound for Québec with provisions; TNA/PRO, ADM 51/940, II, *Success* log, Capt. Rous; and ADM 52/1092, iii, *Vulture* master's log, Mr. Joseph Marriott; and ACCB, I.2, No. 41, statement of Captain Sansin Halsouet, *Ste. Catherine*, 9 October 1755.

[24]ACCB, I.2, No. 48, affidavit submitted to the Chamber of Commerce of Bayonne, 17 March 1757, describing the events of August 1755 at Codroy.

au Choix, where they found French fishing vessels to take them home.[25] Their ships' cargoes of fish, oil and cod roe were taken as prize and delivered to the British authorities at Halifax.[26] As for the residents of the French fishing posts, those who were unable to escape in shallops were herded onto the warships and carried to Île Royale – sixty-seven in total, "most of them women and children," according to the report subsequently filed by Admiral Boscawen – where they were unceremoniously deposited on the shores of Gabarus Bay.[27]

As military operations go, this one had not been very difficult. The British warships encountered no resistance, for neither the French ships they seized nor the inhabitants they expelled were inclined or equipped to challenge them. Indeed, the French had been caught completely off guard. Even when a sudden storm sprang up before operations at Codroy could commence, forcing *Success* to cut its cable and make for the safety of the open sea, the crew of *Deux Maries* could not make their ship ready for sea and escape before *Success* returned. Nor could the French believe, until the moment their ships, gear and fish began to be seized or destroyed, that this was the intent of the English. After all, France and England were nominally still at peace. Captain Rous, however, justified his actions not as acts of war but on the grounds that the French were in violation of the Treaty of Utrecht, which confined the French fishery to the coast north of Point Riche. The protests of Captain La-Freche – that he had been fishing at Codroy for thirty years without any objections from the British – were ignored.[28]

One can only speculate whether Captain LaFreche found any irony in the fact that for years the Newfoundland station ships had been willing to ignore French activity in the area while the naval officers stationed in Nova Scotia, whose jurisdiction did not extend to Newfoundland, were so swift to act. The answer presumably rests with Vice-Admiral Boscawen, who had clear instructions to engage in hostilities against French warships and merchantmen transporting troops or military stores to North America, even though war had

---

[25] ACCB, I.2, No. 40, statement of Jean Lafreche, 9 October 1755.

[26] A claim was subsequently filed for nearly 200,000 *livres* for the loss of the three fishing vessels and cargoes, the boats and shore facilities, and the partial cargo and gear abandoned by *Notre Dame du Rozaire*; ACCB, I.2, No. 48, affidavit submitted to the Chamber of Commerce of Bayonne, 17 March 1757.

[27] TNA/PRO ADM 1/481, 68, Boscawen (*Torbay*, St. Helen's) to the Admiralty, 15 November 1755. No indication has yet been found in the French records that a group of refugees suddenly appeared near Louisbourg; this is one of the areas where additional research is needed.

[28] ACCB, I.2, No. 40, statement of Jean Lafreche, 9 October 1755.

not yet been declared.[29] Rous' actions against French shipping and fishing operations in southwestern Newfoundland were perfectly consistent with Boscawen's orders. At least two of the vessels seized at sea by the ships under Rous' command were carrying provisions to Louisbourg or Québec. Moreover, Rous not only had already engaged in hostile operations against the French at Fort Beauséjour earlier that year but also had been a party to the decision to expel the Acadians from Nova Scotia. That decision had yet to be carried out at the time of the attack on Codroy, but it was driven by a strong sense of insecurity that prevailed among the military authorities in Nova Scotia. The actions against the French settlements and fishing operations in Newfoundland were therefore consistent with the much more massive Acadian deportation that soon followed.

Indeed, one might even be forgiven for wondering whether the eviction of the inhabitants of Port aux Basques and Codroy was not to some degree a dry run for the Acadian deportation. When that much larger eviction took place, John Rous would once again be at the centre of events – it would be the frigate *Success* under his command that would convoy the transports carrying deported Acadians out of the Bay of Fundy on their way to destinations as far away as Georgia and the Carolinas.[30] Had his performance in Newfoundland two months previous given him experience that made him a logical choice for the logistical challenge of deporting the Acadians? Had his record of service against the French fisheries and shipping in Atlantic colonial waters over the course of more than ten years made him the obvious man to choose whenever such an operation was proposed? Was his knowledge of waters in a part of the world where the Royal Navy had not yet established a confident familiarity the reason that the name of John Rous kept appearing in the record of naval operations in Nova Scotia and Newfoundland during the 1740s and 1750s?[31] The answers to these questions must await further research into the career of this remarkably energetic colonial in the British naval service.

---

[29]"...his orders ran: 'If you fall in with any French ships of war or other vessels having on board troops or warlike stores you will do your best to take possession of them. In case resistance should be made, you will employ the means at your disposal to capture and destroy them.'" Corbett, *England in the Seven Years' War*, I, 45.

[30]Douglas, "John Rous," in *DCB*, III, 574.

[31]In 1756, *Success*, still under the command of John Rous, and HMS *Norwich* would be sent by Admiral Charles Holmes first to Bay Vert on the Nova Scotia coast opposite Isle Saint Jean (Prince Edward Island) to check French designs there and then to Newfoundland "to destroy the French fishery there." TNA/PRO, ADM 1/481, 226-227, Charles Holmes, HMS *Grafton*, to Lord Loudon, 12 September 1756.

The French Raid upon the Newfoundland Fishery in 1762: A Study in the Nature and Limits of Eighteenth-Century Sea Power[1]

On 8 May 1762, four French warships slipped out of Brest in thick weather and eluded blockading British cruisers. On board were 560 soldiers and officers under the command of Joseph-Louis-Bernard de Cléron, comte d'Haussonville.[2] Most of the men believed that their destination was Saint-Domingue, France's only remaining possession in the West Indies, the rest having been captured by British forces. Not until they had reached the open sea did the expedition's commander, *Chef d'escadre* Charles-Henri-Louis d'Arsac, chevalier de Ternay, reveal their true objective: the destruction of British trade and commerce in the North Atlantic, beginning with the British fishery at Newfoundland.

The choice of this objective said a great deal about France's situation by that point in the Seven Years' War. For over four years, it had suffered a nearly unremitting series of defeats which left much of its colonial empire in enemy hands, its navy in a shattered condition and its financial and material ability to continue the war very much impaired. Yet attempts to negotiate an end to the war had ended in failure; when the government of William Pitt had refused to satisfy the minimum French demands concerning the fishery in the Gulf of St. Lawrence and at Newfoundland, the French broke off negotiations. That had been in September 1761. The duc de Choiseul, who was then minister of marine, of war and of foreign affairs, had to find some way of pressuring the British into a more reasonable attitude. On the diplomatic side, this would mean bringing Spain (with whom France had concluded an alliance in August) into the war. On the strategic side, it would mean continuing operations against England's maritime empire. Since France's ability to conduct operations was by then quite limited, the campaign for 1762 would have to be planned with careful attention to the selection of the objectives, the organiza-

---

[1]This essay appeared originally in William B. Cogar (ed.), *Naval History: The Seventh Symposium of the U.S. Naval Academy* (Wilmington, DE, 1988), 35-54.

[2]Library and Archives Canada (LAC), Manuscript Group (MG) 2, transcripts, 27-28, copied from France, Archives de la Marine (AM), B4, vol. 104, addendum to Ternay to Choiseul, 5 May 1762. The force which would eventually occupy St. John's was substantially larger, having been augmented by about 300 marines from the ships; LAC, MG 13, copied from Great Britain, National Archives (TNA/PRO), War Office (WO) 34/42, fols. 98-99, Colvill to Sir Jeffery Amherst, 16 August 1762.

tion of the expedition and the appointment of the officers whose responsibility it would be to carry out the mission.

The *Plan de Campagne par Mer pour l'Annee,* 1762, which Choiseul read and endorsed during the winter of 1761-1762, probably provided the inspiration for Ternay's mission.[3] It recommended raids against Brazil (a possession of England's ally, Portugal), the British fishery at Newfoundland and British trade along the European and West African coasts. The plan was never fully adopted; France did not have the naval resources to undertake so ambitious an assault on the British maritime empire. Nevertheless, the object of the proposed campaign, "to make the most of these forces and keep the enemy as busy as possible," was exactly what Choiseul had in mind. By combining some aspects of the plan and discarding others, he was able to prepare a campaign which was much more consistent with French resources and aims.

Choiseul's high regard for the migratory fishery was already a matter of fact; it had been his insistence that French access to the North American fisheries be a *sine qua non* to any peace settlement which had contributed to the breakdown of the peace negotiations in 1761. Both France and England regarded the migratory fishery as an economic resource of the first order.[4] Both countries also valued the fishery as a "nursery for seamen." According to the conventional wisdom of the day, the thousands of landsmen who found annual employment in the fishery became skilled seamen and mariners, thus providing their homelands with a strategic reserve of experienced manpower whenever the onset of war demanded the rapid expansion and deployment of

---

[3]LAC, transcripts, 3-9, from AM, B4, vol. 104.

[4]Ralph G. Lounsbury, *The British Fishery at Newfoundland, 1634-1763* (New Haven, 1934; reprint, New York, 1969), 311-313; and LAC, MG 5, transcripts, 143-146, copied from Archives des Affaires Etrangères, Paris, B[1]/22, "mémoire sur la pesche de L'amérique septentrionale," n.d. [1760-1762]. In fact, both the French and English governments made virtually identical and simultaneous assessments of the value of the migratory fishery. Comparing the resources of Louisiana, Canada and Newfoundland, the British Board of Trade in 1761 concluded that "the Newfoundland Fishery as a means of wealth and power is of more worth than both of the aforementioned provinces," while during the negotiations which finally brought the war to an end, Choiseul maintained that "the cod fishery in the Gulf of St. Lawrence is worth infinitely more for the realm of France than Canada and Louisiana." Board of Trade, cited in Gerald S. Graham, "Fisheries and Sea Power," Canadian Historical Association, *Annual Report 1941*, reprinted in G.A. Rawlyk (ed.), *Historical Essays on the Atlantic Provinces* (Toronto, 1967), 8; and Choiseul, cited in Max Savelle, *The Origins of American Diplomacy; The International History of Anglo-America, 1492-1783* (New York, 1967), 475n.

the national navies.[5] In an age of rival mercantile empires, a country's response to a crisis could well be determined by the war-readiness of its fleets, which in turn could depend upon the availability of trained seamen.[6] It was this dual role as an economic and strategic resource that made the fishery a particularly appropriate target. On the assumption that it represented one of France's better options to bring the war to an end, Choiseul prepared to attack the British fishery at Newfoundland.

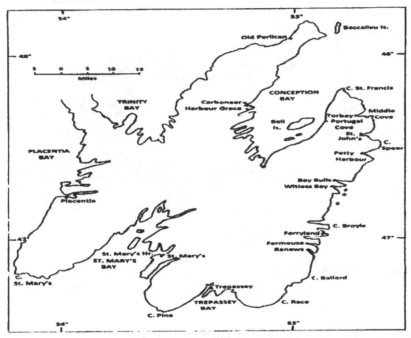

Figure 1:   The Avalon Peninsula

Source: Courtesy of the author.

[5]Keith Matthews, "A History of the West of England-Newfoundland Fisheries" (Unpublished DPhil thesis, Oxford University, 1968), 396; Olaf Janzen, "Newfoundland and British Maritime Strategy during the American Revolution" (Unpublished PhD thesis, Queen's University, 1983), chap. 1; and Jean-François Brière, "Pêche et politique à Terre-Neuve au XVIIIe siècle: La France véritable gagnante du traité d'Utrecht?" *Canadian Historical Review*, LXIV, No. 2 (1983), 168-169.

[6]In June 1770, the North administration was slow to respond to a dispute with Spain over the Falkland Islands, in part because the start of the fishing season had effectively removed thousands of men needed to put the Royal Navy on a war footing. See Nicholas Tracy, "The Falkland Islands Crisis of 1770; Use of Naval Force," *English Historical Review*, XC, No. 354 (1975), 49.

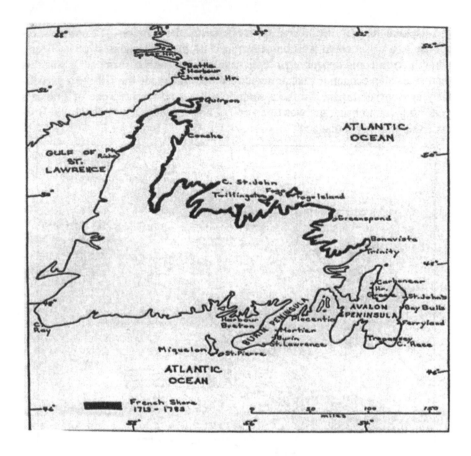

Figure 2:   Newfoundland and the French Shore, 1713-1783

Source: Courtesy of the author.

The object of the mission, according to Ternay's instructions, was "to ravage and destroy, as much as he could, the commerce of the English fishery at the island and on the banks of Newfoundland." Ternay was to attempt the capture of St. John's, which served the fishery at Newfoundland as a rendez-vous, refuge (by virtue of its defences) and administrative centre. He was to proceed with the destruction of that fishery, using St. John's as a base. These operations were to occupy Ternay for no more than a month ("*Ces opérations ne pouvant pas retenir Le S'r. Ch'er. de Ternay plus d'un mois sur ses par-ages*"). He was then to use his discretion in selecting further targets before returning to Brest in the fall ("*Sa Majesté se remet à Son Zèle pour son service d'employer le reste de sa campagne, qui ne droit finir qu'au commencement de Novembre*"). The fishery and commerce at Île Royale were suggested as possibilities, provided that the British forces there were weak. Otherwise, he might return to Europe immediately in order to raid the seaports of Scotland and Ireland. In any case, Ternay was always to destroy enemy shipping, commercial buildings and government structures while exacting wealth from the towns and villages on pain of pillage should they refuse (*"il brusquera toutes les descentes qu'l fera en brûlant tous les Bâtimens qu'jl trouvera dans les Ports, Rades et anses, ainsy que les Magazins et autres Bâtimens civils appartenant à La Marine et au Commerce, et à L'égard des Villes, Bourgs et Villages, il tâchera d'en tirer les plus fortes contributions sous peine de pillage en cas de refus"*).

Finally, Ternay's instructions included a sentence which granted him virtually complete freedom to select other targets should he think that they would better serve the object of his mission (*"Sa Majesté...lui laisse une entière liberté sur le chois des objets auxquels il croira devoir s'attacher de préférence pour achever sa campagne"*). Such a concession was not uncommon in the eighteenth century. It left a good commander free to respond to an unexpected contingency or opportunity. There was always the danger, however, that a weak or inexperienced commander would abuse that freedom to limit or even to abandon the mission altogether. In Ternay's case, this is precisely what would happen.[7]

While there are some who view Ternay's expedition as an attempt to capture and secure a bargaining counter with which France could negotiate a more equitable peace, it is clear that the attack on the fishery was planned as a raid.[8] Indeed, the successful outcome of the mission was assured only if Ter-

---

[7]LAC, MG 2, transcripts, copied from AM, B⁴/1041, "mémoire du Roy pour Servir d'Instruction au S'er. Ch'er. de Ternay."

[8]The French attack upon the fishery in 1762 is identified as a raid in Julian S. Corbett, *England in the Seven Years' War: A Study in Combined Strategy* (2 vols., London, 1907; 2nd ed., London, 1918; reprint, New York, 1973), II, 323; and in Maurice Linÿer de la Barbée, *Le Chevalier de Ternay* (Grenoble, 1972), chap. 8. It is identified as a conquest by Daniel Woodley Prowse, *A History of Newfoundland from the English Colonial and Foreign Records* (London, 1895; reprint, Belleville, ON, 1972), 305; Gordon O. Rothney, "The History of Newfoundland and Labrador, 1754-1783" (Unpublished MA thesis, University of London, 1934), 49; David Webber, "The Recapture of St. John's, 1762," *Newfoundland Quarterly*, LXI, No. 3 (1962), 1-10; Glanville Davies, "England and Newfoundland: Policy and Trade, 1660-1783" (Unpublished PhD thesis, University of Southampton, 1980), 122; and Bernard Ransom, "A Century of Armed Conflict in Newfoundland," *Newfoundland Museum Notes*, No. 10 (St. John's, 1982). Gerald Graham refers to the attack as both a raid and an attempt "to win one substantial bargaining counter," which implicitly would mean a conquest; Gerald S. Graham, *Empire of the North Atlantic; The Maritime Struggle for North America* (Toronto, 1950), 191. This is a contradiction since, by definition, "over-sea invasion is a continuous process which...requires the maintenance of unbroken communications during the period necessary to ensure complete military success. A raid, on the

nay avoided conquest. So long as his force continued to move from objective to objective, the British response would have to be a limited one, despite overwhelming superiority in sea power. It was precisely for this reason that Ternay was nervous about seizing and occupying St. John's, even for only a month; as he later explained to Choiseul, "I do not see how it would be possible to hold a post which is surrounded by the enemy and far from all help."[9] But given the slow rate of speed with which news of his presence at St. John's could be transmitted in the age of sail, a stay of one month ought not to have jeopardized the success of the mission. Apparently Choiseul had recognized what Julian Corbett would later articulate, namely, that "command of the sea means nothing but control of sea communications."[10] So long as Ternay did not remain too long, his precise location would remain unknown, and any British response to his presence in the North Atlantic would be unlikely to intercept him.

Certainly by the spring of 1762, the benefits to be won outweighed the risks; by then it was apparent that Choiseul's diplomatic initiative had failed. The alliance with Spain brought that country into the war in January 1762. Choiseul had been encouraged by the prospect of adding Spain's naval resources to those of France, even as England's Royal Navy would be taxed by the need to contain this new threat. Moreover, Spain's entry into the war would terminate England's lucrative commercial relations with that country, thereby increasing the strain on England's ability to finance the war. Finally, Spain's participation would threaten England's ally, Portugal, and force the British government to disperse its military resources even further.

But Spain's entry into the war proved to be a disappointment. Hostilities against Portugal did not begin until May 1762, by which time British forces had seized Martinique, leaving St. Domingue the last remaining French possession of significance in the Caribbean. The capture of Martinique in February was immediately followed in March by the commencement of operations in Cuba. Choiseul was quite willing to resume negotiations at this point, particularly since the hated Pitt had resigned in October 1761 and had been replaced by the Earl of Egremont. Nevertheless, Choiseul "was too able a statesman to abandon his plans for war simply because of the apparently favourable

---

other hand, is an attempt to seize by surprise some strong point on or close to a coastline, with a view to inflict an injury – moral or material – which might cripple, or at least diminish, the fighting resources of an enemy." Sir George Sydenham Clarke, *Fortification: Its Past Achievements, Recent Development, and Future Progress* (London, 1907; reprint, Charleston, SC, 2012), 170-171.

[9]LAC, MG 2, transcripts, 37, copied from AM, B⁴/104, Ternay to Choiseul, 9 July 1762.

[10]Corbett, *England in the Seven Years' War*, I, 308.

attitude of the British Cabinet towards peace."[11] The attack upon the British fishery proceeded even as negotiations between France and England resumed.

Ternay's expedition therefore must have served Choiseul's purposes well. Not only did it make use of what few ships were still at his disposal, but it also might even lead to dividends at the peace table – and at relatively modest cost and risk. Nevertheless, any success which might ensue would depend not so much on the logic of the expedition as upon the men responsible for its execution. The military contingent which embarked at Brest was as powerful as space on the warships would allow; according to later British reports, the soldiers were all "Choice men."[12] Yet the crews of the four warships in Ternay's squadron – the seventy-four-gun *Robuste*, sixty-four-gun *Eveillé*, thirty-gun frigate *Garonne* and twenty-six-gun storeship *Licorne* – were sickly and understrength.[13] Aggressive action would be avoided as a result, and Ternay's leadership would be uncertain. Moreover, he had only recently been promoted to the rank of ship's captain, and he had never commanded a ship in action, let alone a squadron entrusted with the complex task of carrying out a combined operation. On the other hand, he had recently distinguished himself for his role in bringing the warships *Robuste* and *Eveillé* out of the Vilaine River to Brest. They had been driven over the bar at the river's mouth by pursuing British warships during the Battle of Quiberon Bay in 1759 and had idled there ever since, secure but useless.[14] For this action Ternay received his captaincy and the command of the Newfoundland expedition. Evidently, for such a mission Choiseul valued audacity and initiative more than he did experience.

---

[11]Savelle, *Origins of American Diplomacy*, 497.

[12]LAC, MG 13, copied from TNA/PRO, WO 34/42, fols. 98-99, Colvill to Amherst, 16 August 1762.

[13]Ternay had difficulty bringing his crews up to strength despite stripping men from the ships still trapped in the Vilaine River; LAC, MG 2, transcripts, 29-30, copied from AM, B⁴, vol. 104, Ternay to Choiseul, 7 April and 5 May 1762. Few of the men had been to sea, and most were dressed in rags, according to Linÿer de la Barbée, who described them as "a wretched herd which was therefore assured a frightening mortality rate;" Linÿer de la Barbée, *Le Chevalier de Ternay*, 155. *Biche*, which some sources add as a fifth vessel to Ternay's squadron, was in fact an English privateer taken as Ternay returned to European waters following the failure of his expedition; Linÿer de la Barbée, *Le Chevalier de Ternay*, 169-170.

[14]France, Bibliothèque Nationale, Département des Manuscrits, Nouvelles acquisitions françaises 9410, "Journal et Lettres du Chevalier de Ternay," 273-73v, La Pérouse to Ternay, 1 February 1762. The entire operation is described in Linÿer de la Barbée, *Chevalier de Ternay*, chap. 7.

While success therefore seemed possible, Ternay cautioned that "only secrecy and extreme diligence can assure a project's success."[15] Toward this end, various precautions were taken to disguise the purpose behind the activity at Brest. Mail intended for St. Domingue was sent on board Ternay's ships, as were two or three officers attached to regiments serving there. Ternay also was to send dispatches from Newfoundland which implied that St. Domingue was still his ultimate destination. Finally, he was instructed to chart a course for Newfoundland which would minimize the risk of being seen or of revealing his true objective; should he encounter any enemy merchantmen, he was to take or destroy them.[16]

Most of these measures had the desired effect. The London newspapers were aware that something was up, but they could only speculate about Ternay's destination. And the British did not immediately detect his departure.[17] But in one of those examples of bad luck for which no amount of careful planning can allow, Ternay's squadron was barely three days out of Brest when it crossed the path of a British convoy. The convoy escort formed a disposition for battle, but the French, who were undermanned and had one ship *en flûte*, were in no mood to risk their mission for a battle whose outcome was uncertain. Consequently, the French squadron tacked to the northward and escaped.[18] Although Ternay's mission would continue to remain a mystery, the British Admiralty was now alerted to his presence in the North Atlantic.[19]

The encounter left Ternay worried that British warships assigned to the Newfoundland station would soon be responding to his presence there. As the French squadron approached the Grand Banks, another incident intensified his concern. *Garonne* became separated from the squadron and, contrary to

---

[15]LAC, MG 2, transcripts, 29, copied from AM, B⁴/104, Ternay to Choiseul, 7 April 1762.

[16]*Ibid.*, transcripts, 18-21, Ternay's instructions, enclosed in Choiseul to Ternay, 30 April 1762; and transcripts, 22-23, Choiseul to Ternay, 9 April 1762.

[17]"[They] are the very armament the Papers mention'd before we left England," LAC, MG 12, microfilm, copied from TNA/PRO, Admiralty Papers (ADM) 1/1835, part 3, Graves to Clevland, 21 September 1762.

[18]John Campbell, *Lives of the British Admirals* (8 vols., London, 1742-1744; rev. ed., London, 1812-1817; reprint, Charleston, SC, 2012), V, 192-193; and William L. Clowes, *The Royal Navy: A History from the Earliest Times to the Present* (7 vols., London, 1897-1903; reprint, Charleston, SC, 2012), III, 250-251.

[19]On 27 June 1762, orders were sent to Captain Graves, the senior officer on the Newfoundland station, "acquainting [him] with Mons'r Ternays force, and directing [him] how to act if he came to Newfoundland." LAC, MG 12, copied from TNA/PRO, ADM 1/1835, part 3, Graves to Clevland, 21 September 1762.

instructions, attempted unsuccessfully to capture an English vessel. The merchantman was last seen making for St. John's, and Ternay was certain that not only the fishery but also the naval establishment at Halifax, Nova Scotia, would know that he was loose in Newfoundland waters. It was therefore with a sense of extreme urgency that Ternay brought his ships into Bay Bulls, a few miles south of St. John's, on 24 June. D'Haussonville and the troops were quickly landed with instructions to march upon St. John's – an arduous task since there were no roads, only primitive wilderness paths; they were expected to complete the trek in three days. Ternay would remain at Bay Bulls, destroying the shipping and fishing facilities found there. By descending upon St. John's from the rear instead of from the sea as expected, he hoped to salvage some of the element of surprise which he believed had been compromised.[20]

Ternay did not yet realize just how vulnerable the Newfoundland fishery was to a raid at that time. Despite the high regard in which the fishery was held, it had been traditional British practice to defend it through the exercise of sea power in metropolitan waters.[21] Little had been done in Newfoundland itself to provide the fishery with an effective local defence. The only significant fortifications on the island were at Placentia and St. John's. These were so poorly situated and in such a decrepit state by 1762 that their ability to repulse an attack was in doubt. Ternay would be shocked by the condition of the defences at St. John's, observing that "they say that this place, which has been in England's possession for a long time, has cost that country £190,000. To see its actual state, you would not think so."[22]

The fortifications were garrisoned by two companies of the Fortieth Regiment of Foot, which had been serving in Newfoundland since 1717, the year in which their parent regiment had been formed; by 1762, the company at Placentia was described as being "very defective," and there is nothing to sug-

---

[20]LAC, MG 2, transcripts, 33-36, copied from AM, B⁴/104, Ternay to Choiseul, 9 July 1762.

[21]Gerald S. Graham, "Britain's Defence of Newfoundland; A Survey from the Discovery to the Present Day," *Canadian Historical Review*, XXIII, No. 3 (1942), 260-279.

[22]LAC, MG 2, transcripts, 33-36, copied from AM, B⁴/104, Ternay to Choiseul, 9 July 1762. The defects of the works at Placentia are described in LAC, MG 11, transcripts, 45-49, copied from TNA/PRO, Colonial Office (CO), 5/161, Report of Leonard Smelt, Engineer in Ordinary, to the Board of Ordnance, 22 November 1751; see also Jean-Pierre Proulx, *Placentia, Newfoundland, The Military History of Placentia: A Study of the French Fortifications* and *Placentia, 1713-1811* (Ottawa, 1979), 143-145. The works of St. John's are described in LAC, MG 12, copied from TNA/PRO, ADM 1/481, fols. 384-386 and 423-424, Captain Richard Edwards to Admiral Francis Holburne, 17 July and 6 September 1757.

gest that the condition of the company at St. John's was any different.[23] A company of the Second Battalion, Royal Regiment of Artillery, also was serving in Newfoundland, but the largest detachment was at Placentia, not St. John's.[24] Finally, the British warships stationed at Newfoundland every year from late spring to early fall for the regulation and defence of the fishery must be considered. In 1762, these were the fifty-gun *Antelope* and two twenty-gun frigates, *Syren* and *Gramont*. At the time of Ternay's appearance off St. John's, only the frigates had arrived from England; *Antelope*, with Captain Thomas Graves, the commander in chief of the Newfoundland station and governor of the island, was still at sea. To all intents and purposes, there was no effective local defence.

This is not to say that some sort of resistance was not possible. *Gramont* had arrived at St. John's a few days earlier with some of the trade and was moored in the harbour. Its commander, Captain Patrick Mouat, was able to send about 100 of its sailors and marines up to Fort William to help serve the guns. At the same time, Captain Lieutenant Rogers was keen to have his artillery detachment open fire upon the French. A large number of fishermen and inhabitants – about 370 men by one report – had volunteered to assist the garrison in the defence of the town, and many were subsequently issued arms and ammunition. But the garrison commander, Captain Walter Ross, would give them no further encouragement or direction, nor would he permit Rogers to begin bombarding the French, except for two ineffectual shots fired while the enemy was still half a mile away. As a result, d'Haussonville advanced his men to within 300 yards of the fort, where they found shelter behind a hill. He then summoned the garrison to surrender. Even then, defeat was not inevitable, for it had been impossible for the French to bring their artillery with them on the difficult march from Bay Bulls; they would have had to take the fort by storm. But Ross believed that the French numbered more than 1500 and that, in the face of such numbers, resistance was futile. Incapable of providing decisive leadership even in surrender, and much to the disgust of the other officers, he called a council of war before agreeing to the French terms. Captain Mouat was particularly bitter because Ross' sudden surrender

---

[23]J.A. Houlding, *Fit for Service: The Training of the British Army, 1715-1795* (Oxford, 1981), 18-19; R.H. Raymond Smythies, *Historical Records of the 40th Regiment (2nd Somerset) Regiment, Now 1st Battalion The Prince of Wales Volunteers (South Lancashire Regiment) from Its Formation in 1717, to 1893* (Devonport, 1894), 34; LAC, MG 11, copied from TNA/PRO, CO 194/15, fols. 31-34, Graves to Board of Trade, with enclosure, 18 August 1762; LAC, MG 12, copied from TNA/PRO, ADM 1/1835, part 3, Graves to Clevland, 18 August 1762; and LAC, MG 12, copied from TNA/PRO, ADM 1/481, fols. 423-424, Edwards to Holburne, 6 September 1757.

[24]M.E.S. Laws, *Battery Records of the Royal Artillery, 1716-1859* (Woolwich, 1952), 30.

had left *Gramont* trapped in the harbour. Mouat had to order its guns spiked and the ship scuttled in an attempt to keep it out of enemy hands.[25]

Immediately after the capitulation of Fort William on 29 June, the French began vigorous efforts to improve the state of defence at St. John's. Michael Gill, a local magistrate, was compelled to provide workmen to improve the fortifications.[26] The captured crew of *Gramont* was put to work to raise the sunken frigate. Its guns were then drilled out, and the ship was prepared as swiftly as possible to be added to the French squadron.[27] Meanwhile, on 3 July, Ternay had brought his ships into the harbour and then secured its entrance with a boom and a new battery.[28] Four sloops and a schooner, armed with cannon and mortars and loaded with 150 soldiers, were sent north to destroy the defences and fisheries of Conception, Trinity and Bonavista bays. Two other vessels were sent to the Southern Shore on a similar mission.[29] M. de Clonard, who was directed to raise recruits for a marine regiment, had found numerous Irishmen willing to enter into the French service.[30] So far, the mission was proceeding much more smoothly than Ternay had anticipated.

Yet as success followed success, Ternay became increasingly reluctant to proceed with his mission by attacking the North American fishery. His instructions had directed him to stay in Newfoundland no longer than one

---

[25]Account compiled from information provided by soldiers on board the brig *Dorothy*, a cartel carrying prisoners from Newfoundland to England, in LAC, MG 13, copied from TNA/PRO, WO 34/26, fols. 202-203, enclosed in Governor Francis Bernard to General Jeffery Amherst, 29 August 1762; LAC, MG 13, copied from TNA/PRO, WO 34/18, fols. 53-53v, "An Account of the Enemy Fleet and Army at Newfoundland as Given by Lt. Fade to Col. Tulleken," 29 July 1762; LAC, MG 11, transcripts, 477-480, copied from TNA/PRO, CO 5/62, statement of Captain Lamb, 1 August 1762; LAC, MG 12, copied from TNA/PRO, ADM 1/482, fols. 210, Colvill to Clevland, 24 July 1762; and LAC, MG 12, copied from TNA/PRO, ADM 1/2115, Captain Mouat to Clevland, 28 February 1763.

[26]LAC, MG 11, transcripts, 477-478, copied from TNA/PRO, CO 5/62, Statement of Captain Lamb, 1 August 1762.

[27]LAC, MG 2, transcripts, 34-35, copied from AM, B⁴/104, Ternay to Choiseul, 9 July 1762.

[28]*Ibid.*, transcript, 34.

[29]*Ibid.*, transcript, 40; and transcripts, 42-49, "Relation d'une Action de Guerre qu'a fait Le Ch'r. De La Motte Vauvert...en 1762," 25 February 1775.

[30]*Ibid.*, transcripts, 24-25, Choiseul to Ternay, 12 April 1762; transcript, 41, Ternay to Choiseul, 9 July 1762; and LAC, MG 4, transcript, 147, copied from Archives de la Guerre, Paris, B1, A1, vol. 3628, de Clonard to Choiseul, 16 October 1762.

month after the capture of St. John's, but within a week of his arrival at that port he had already postponed his departure to the beginning of September.[31] Ternay offered several explanations for the delay; he needed time to complete the improvements to the harbour defences, needed time to careen and repair *Gramont* and was concerned that the British warships stationed at Halifax had been alerted about his presence and were expecting him.[32] Possibly the best explanation came from the British commander at Louisbourg, Lieutenant Colonel Tulleken. He was confident that the French would make no further attempts upon British possessions in the region. "It seems to me," he wrote to Sir Jeffery Amherst, his commander in chief, "that what they have got has been more fortunate than from any Expectations they would have had of success; and finding themselves so lucky, they endeavour to fix where they are in hopes of being speedily reinforced from Home." In effect, the French were stunned into immobility by their own success.[33]

It was a perceptive observation. Before sailing from Brest, Ternay had questioned whether St. John's should or even could be taken.[34] When, contrary to his expectations, St. John's fell with scarcely a shot being fired, he tried lamely to cover his surprise by declaring that "in war it is usually the most difficult tasks which succeed most easily."[35] Nevertheless, he was unwilling to test this observation any further. Rather than proceed with his attack upon British commerce in the North Atlantic, Ternay decided to remain where he was and prepare St. John's for permanent occupation. It was a decision which ignored a fact which he had recognized before leaving Brest, namely that by 1762, France no longer had the means to support a prolonged occupation of Newfoundland.

It was also a decision which ignored the facts of Ternay's own experience. Despite having landed over 500 troops with ease at Bay Bulls, he now maintained that the entire coast was too sheer to be accessible even by shallops. Despite d'Haussonville's success at marching his men twenty-five miles

---

[31]LAC, MG 2, transcripts, 18-24, copied from AM, B⁴/104, "mémoire du Roy pour Servir d'Instruction au S'r, Ch'er. de Ternay," enclosed in Choiseul to Ternay, 30 April 1762; and transcript, 39, Ternay to Choiseul, 9 July 1762.

[32]*Ibid.*, transcripts, 39-41; and transcript, 50, Ternay to Choiseul, 20 July 1762.

[33]LAC, MG 13, copied from TNA/PRO, WO 34/18, fol. 49v, Tulleken to Amherst, 20 July 1762.

[34]LAC, MG 2, transcripts, 37-38, copied from AM, B⁴/104, Ternay to Choiseul, 9 July 1762.

[35]*Ibid.*, transcript, 34.

in three days, Ternay now claimed that overland travel was almost impossible owing to the bad terrain and lack of roads. Despite capturing St. John's from the landward side, Ternay assumed that any British countermeasures would come from the sea. With the dust still settling from the French seizure of the town, Ternay now assured Choiseul that St. John's could be made invincible with very little extra work.[36] To Ternay, the security of having a bird in the hand was incalculably more attractive than the risks of seeking more in the bushes.

Having made his decision, Ternay then had to solve the pressing problem of securing the provisions necessary for a prolonged occupation. Troops were sent through the town to survey the supplies in the storehouses of the local merchants and to post sentinels over them. Livestock and poultry were confiscated for garrison use. The detached force which had been sent to destroy the fishery north of St. John's also was ordered to collect provisions for the French garrison; indeed, Ternay regarded that as being "one of the principal objects of his mission."[37] He had arrived shortly after the winter's consumption of food reserves and before the spring trade had arrived with fresh provisions and supplies.[38] A few vessels did arrive after the French had taken St. John's; these were seized as they entered the harbour, before they realized the true situation.[39] Most of the trade, however, was warned in time and either turned back or headed elsewhere.

Unless measures were taken to reduce the number of mouths to be fed, Ternay simply could not expect his force to hold St. John's until relief could be sent from France. Consequently, most of the inhabitants, except for a few people who had value as hostages, were loaded onto captured merchantmen and allowed to make their way as best they could to England or the American colonies.[40] Only the Irish Catholics who were sympathetic to the

[36]*Ibid.*, transcripts, 37-39.

[37]*Ibid.*, transcript, 40; transcript, 42, "Relation d'une Action de Guerre qu'a fait Le Ch'r, De La Motte Vauvert...en 1762;" LAC, MG 11, transcripts, 477-478, copied from TNA/PRO, CO 5/62, statement of Captain Thomas Lamb, 1 August 1762; and LAC, MG 17, B1, transcripts, 44-45, copied from Society for the Propagation of the Gospel in Foreign Parts, London (SPG), Letters 1700-1786, B/6, Rev. Edward Langman (St. John's) to Rev. Secretary Daniel Burton, 14 July 1762.

[38]LAC, MG 17, B1, transcripts, 46-47, copied from SPG, B/6, Langman to Burton, 2 November 1762.

[39]*Ibid.*

[40]*Ibid.*; LAC, MG2, transcript, 39, copied from AM, B⁴/104, Ternay to Choiseul, 9 July 1762; and LAC, MG 12, copied from TNA/PRO, ADM 1/482, fols. 223 and 215, Colvill to Clevland, 16 August and 20 September 1762.

French were allowed to stay.[41] The French garrison also would be reduced to 350 men under the command of Lieutenant Colonel de Bellecombe, while the rest of the soldiers would return to France with d'Haussonville and Ternay's squadron, together with as many Irishmen as would enter into French service. In this way the total population of St. John's would be reduced to a maximum of 500 people by the time winter arrived. With a little belt-tightening and the help of supplies and stores which Ternay hoped could be sent from France before the year was out, the provisions would last until early 1763. Provided another squadron were then to bring reinforcements and supplies to Newfoundland, France would be guaranteed possession of St. John's through 1763.[42]

If Choiseul was disturbed by the way in which Ternay had transformed what was to have been a raid into a conquest, he gave no sign of it when the news reached him early in August 1762. Instead, he expressed both his and the king's pleasure and satisfaction with Ternay's achievement.[43] It is possible that Choiseul was reluctant to admit in any way that Ternay had proved to be an unwise choice as commander of the expedition. A more likely explanation for Choiseul's acceptance of Ternay's *fait accompli* is suggested by the peace negotiations between France and England. These had resumed during the winter of 1761-1762 and were being conducted through the intermediation of the Sardinian envoys to Paris and London, the bailli de Solar and Count Viry.

By the summer of 1762, these negotiations had reached a critical stage; Choiseul was anxious to make peace, but France's ally, Spain, was being obtuse and the British cabinet was divided over the terms.[44] Although de Solar believed that further delay in the progress of the peace talks would encourage the French to think seriously about keeping Newfoundland, there is no evidence to suggest that Choiseul tried to use the island as a "bargaining counter," nor does it seem likely, at so critical a juncture in the negotiations,

---

[41]How many Irish Catholics actually were in St. John's in 1762 is not certain. During the winter of 1760-1761, the Irish population was described as being 400 men, 100 women, sixty boys and sixty girls, figures which clearly were meant to be estimates; see LAC, MG 17, transcript, 39, copied from SPG, B/6, Langman to Bearcroft, 4 November 1762.

[42]LAC, MG 2, transcripts, 39-40, copied from AM, B⁴/104, Ternay to Choiseul, 9 July 1762.

[43]LAC, MG 1, transcripts, 59-63, copied from Archives des Colonies, Paris (AC), B/104, Choiseul to Ternay, 14 August 1762.

[44]Savelle, *Origins of American Diplomacy*, 489-510; Ronald Hyam, "Imperial Interests and the Peace of Paris (1763)," in Hyam and Ged Martin (eds.), *Reappraisals in British Imperial History* (Toronto, 1975), 21-43; and Zenab Esmat Rashed, *The Peace of Paris, 1763* (Liverpool, 1951), chaps. 4-5.

that Choiseul would have jeopardized a settlement by assuming a position which seemed certain to prolong the war.[45] Nevertheless, he did agree to send two supply ships to St. John's before winter set in.[46] Ternay may not have adhered to his instructions, but his success at St. John's had given France a momentary moral advantage which Choiseul would be anxious to preserve, at least until the terms of the peace were worked out.

The questions which now had to be faced were how quickly and how effectively the British would respond to the capture of St. John's. Ternay and d'Haussonville were confident that the British would be able to do no more than send some warships to Newfoundland which might menace the French but not dislodge them. This was why Ternay saw no contradiction in improving only the seaward defences of St. John's, despite having taken the town from the landward side. Similarly, when de Cirq, the French engineer, encouraged d'Haussonville to entrench upon the hills which commanded Fort William, his advice was ignored.[47] Even Choiseul agreed that a British counterattack with soldiers and artillery seemed out of the question before 1763.[48] And, so long as the British were uncertain about Ternay's intentions, this remained a valid conclusion.

When news of the capture of St. John's reached London around 21 July, the government sent a squadron of warships under Captain Hugh Palliser to Newfoundland. There he was to reinforce Captain Graves who, it was expected, would already have been joined by Lord Colvill and the North America-stationed ships.[49] No troops were sent with Palliser, nor was there any reason to suggest that troops were thought necessary. Meanwhile, HMS *Syren*

---

[45]LAC, MG 23, transcripts, 2, 125-126, copied from William L. Clements Library, Ann Arbor, MI (WLCL), Earl of Shelburne Papers, vol. 10, de Solar to Viry, 26 August 1762.

[46]LAC, MG 1, transcripts, 59-60, copied from AC, B/104, Choiseul to Ternay, 14 August 1762.

[47]LAC, MG 2, transcripts, 38-39, copied from AM, B/104, Ternay to Choiseul, 9 July 1762; and LAC, MG 23, transcripts, 28, 58, copied from WLCL, Shelburne Papers, 86, Captain Hugh Debbieg to Lord Shelburne, 19 February 1767.

[48]LAC, MG 1, transcripts, 64-67, copied from AC, B/104, Choiseul to de Bellecombe, 14 August 1762.

[49]*London Daily Advertiser*, 26 July 1762, cited in W.S. Lewis (ed.), *Horace Walpole's Correspondence, vol. 22: Correspondence with Mann* (New Haven, 1960), 52n-53n. Palliser sailed on 7 August 1762 with three seventy-fours and a thirty-two-gun frigate. He arrived at St. John's on 19 September, too late to participate in the recapture of the town. LAC, MG 12, copied from TNA/PRO, ADM 1/2299, part 8, Palliser to Clevland, 22 August 1762.

came into Halifax on 30 June with the news that enemy ships were in the vicinity. The frigate was part of Graves' command and had been patrolling the fishery out of Placentia when it spoke to a schooner which claimed to have encountered five enemy ships, "Suppose to be Spanish Men of War."[50]

Lord Colvill received the report with skepticism and suggested that the schooner had probably seen some enemy frigates or privateers, but he was prepared to cruise with his ships toward Newfoundland "to take, or Drive them off the Coast."[51] But the civil authorities at Halifax were convinced not only that the report was accurate but also that their colony, Nova Scotia, was the enemy's real objective, and they refused to let Colvill sail.[52] Martial law was declared, an embargo was imposed on all shipping and a militia company was hurriedly placed on daily guard duty.[53] No one suggested that a military force be assembled for any purpose other than defending Nova Scotia from the expected attack.

Throughout the month of July, British military and naval authorities in North America remained alert to the possible presence of an enemy force in their vicinity, but without more precise information they could not act. At Halifax, Lord Colvill was prepared to lead an expedition to Newfoundland once it became apparent, late in the month, that the fishery there was the enemy's main objective. But without orders from General Amherst in New York, the military commanders at Halifax and Louisbourg refused to provide Colvill with troops.[54] Amherst had been puzzled by the reports he was receiving and had adopted a wait-and-see attitude. Some of the reports continued to describe

---

[50]LAC, MG 11, transcripts, 412-413, copied from TNA/PRO, CO 5/62, Major General Sir Jeffery Amherst to Rear Admiral Rodney, 20 July 1762; LAC, MG 12, copied from TNA/PRO, ADM 1/1835, part 3, Captain Thomas Graves to Clevland, 18 August 1762; and LAC, MG 11, copied from TNA/PRO, CO 194/15, fols. 31-34, Graves to Board of Trade, 18 August 1762.

[51]LAC, MG 11, transcripts, 412-413, copied from TNA/PRO, CO 5/62, Amherst to Rodney, 20 July 1762.

[52]John Bartlett Brebner, *The Neutral Yankees of Nova Scotia: A Marginal Colony during the Revolutionary Years* (New York, 1937; reprint, Toronto, 1969), 37-39; and LAC, MG 12, copied from TNA/PRO, ADM 1/482, fol. 208, Colvill to Clevland, 2 July 1762.

[53]Philip Chadwick Foster Smith (ed.), *The Journals of Ashley Bowen (1728-1813) of Marblehead, vol. 1* (Boston, 1973), 133; and letter of 15 July 1762 from Halifax and printed 29 July 1762 in the *Boston News-Letter and New England Chronicle*, 133n.

[54]LAC, MG 12, copied from TNA/PRO, ADM 1/482, fols. 211-213, Colvill to Clevland, 30 July and 6 August 1762.

the enemy as Spanish; this made little sense since, at that moment, British forces were attacking Cuba, and one would therefore expect all available Spanish warships to be sent to the Caribbean.[55] One report implied that there was in fact no enemy force, only "some Victuallers bound for Quebec," which may inexplicably have fired upon some fishing vessels.[56] Consequently, Amherst would only caution the garrisons in Nova Scotia, Louisbourg and Québec to be on their guard.[57]

Even after 20 July, when he received confirmation about Ternay's attack upon the fishery and the capture of St. John's, Amherst would not initiate a counterattack. He still believed that the assault on Newfoundland was only a raid, which would be followed by another upon one or more of the North American colonies. Amherst was all too conscious of his weakness in such an event. Most of the soldiers stationed in North America were serving in the campaign against Cuba; according to one New York merchant, "the Troops are so drained that General Amherst has not so much as a sentinel."[58] Of even greater concern to the general was the fact that the same campaign had taken nearly every ship belonging to the North American station; in a letter to Rear Admiral George Rodney, Amherst revealed that "this coast...is Entirely bare from Newfoundland to Florida."[59] Consequently, when he began to seek reinforcements, it was warships that he requested rather than troops.[60] Although Amherst did order all the transport vessels then in New York into a state of readiness, this must be seen as a general measure, taken for "the pro-

---

[55]LAC, MG 11, transcripts, 420-421, copied from TNA/PRO, CO 5/62, statement of Ganeston Meers, 11 July 1762, enclosed with Amherst to Rear Admiral Rodney, 20 July 1762.

[56]*Ibid.*, transcripts, 422-423, statement of Samuel Dogget, 11 July 1762; and transcript, 416, Amherst to Admiral Sir George Pocock, 20 July 1762.

[57]*Ibid.*, transcripts, 306-307, Amherst to Secretary of State Egremont, 20 July 1762; and J. Clarence Webster (ed.), *The Journal of Jeffery Amherst, Recording the Military Career of General Amherst in America from 1758-1763* (Toronto, 1931), 287.

[58]Cited in John Shy, *Toward Lexington: The Role of the British Army in the Coming of the American Revolution* (Princeton, 1965), 106.

[59]LAC, MG 11, transcript, 412, copied from TNA/PRO, CO 5/6, Amherst to Rodney, 20 July 1762.

[60]See *ibid.*, transcripts, 304-305 and 409-419, Amherst's letters of 20 July 1762 to Secretary of State Egremont, General Lord Albemarle, Admiral Pocock and Rear Admiral Rodney. Pocock's response can be found in his letter of 9 October 1762 to Clevland in David Syrett (ed.), *The Siege and Capture of Havana, 1762* (London, 1970), 299-300.

tection of every part of this Continent," and not as evidence that he was already organizing an expedition for the relief of St. John's.[61] Until he knew for certain where the French would be, Amherst could only wait and prepare.

Ternay thus had been proven correct, for he had predicted that the British would do nothing until they knew with certainty where the French were.[62] But he was quite mistaken in assuming that, once the location of the French had been determined, the British would not organize a counterattack before 1763. On the contrary, when General Amherst learned on 8 August that "it appears the Enemy intend to Keep possession of that place [St. John's]," he acted with impressive speed and force.[63] Amherst correctly assumed that the British government would probably send warships, but not troops, to assist in the recapture of St. John's. He also recognized, now that the danger of further attacks by Ternay upon North America had been dispelled, that the colonies could be safely stripped of their defences in order to assemble a counterstroke.

Amherst had no troops of his own, but he was able to scrape together about 200 regulars out of the hundreds convalescing at New York from wounds and sickness contracted during the campaign against Martinique.[64] These men would form the nucleus of the expedition, and Amherst's brother, Lieutenant Colonel William Amherst, would be its commander. The transports were quickly loaded and sailed on 14 August for Halifax and Louisbourg where they received another 1300 regulars, provincials and artillerymen before continuing on to Newfoundland. On 10 September, the transports made Cape Race. Barely a month had elapsed since General Amherst had begun to organize the expedition.[65]

It was a remarkable accomplishment, not only for the speed with which the expedition had been organized but also for the way in which it demonstrated how well General Amherst, like Choiseul, had grasped that "com-

---

[61]LAC, MG 11, transcript, 453, copied from CO 5/62, Amherst to Colvill, 9 August 1762; and transcripts, 306-307, Amherst to Egremont, 20 July 1762.

[62]LAC, MG 2, transcript, 37, copied from AM, B4, vol. 104, Ternay to Choiseul, 9 July 1762.

[63]LAC, MG 11, transcript, 427, copied from TNA/PRO, CO 5/62, Amherst to Egremont, 15 August 1762.

[64]*Ibid.*, transcripts, 427-428; transcripts, 436-440, Amherst to William Amherst, 13 August 1762; and transcript, 445, "List of the Troops Ordered to Form the Corps under the Command of Lt. Colonel Amherst," 13 August 1762.

[65]*Ibid.*, transcripts, 527-529, William Amherst to General Amherst, 27 August 1762; and J.C. Webster (ed.), *The Recapture of St. John's, Newfoundland in 1762 as Described in the Journal of Lieut.-Colonel William Amherst, Commander of the British Expeditionary Force* (Shediac, NB, 1928), 5-6.

mand of the sea means nothing but control of sea communications."[66] Amherst had recognized that the French decision to remain at St. John's had changed the complexion of the situation completely. By abandoning any hope or pretence of threatening their sea communications, Ternay had surrendered the initiative to the British. Had he continued to attack the British trade in North American waters, Amherst would have been forced to proceed much more cautiously with his preparations for the recovery of St. John's; Captain Graves and Lord Colvill barely had sufficient ships between them to face Ternay in battle, let alone be able to spare any to escort Colonel Amherst's transports. But by remaining at St. John's, Ternay allowed the transports to proceed from New York to Halifax, then to Louisbourg and finally on to Newfoundland without any escort whatsoever. Although the task of dislodging the French from their improved defences at St. John's still lay ahead, to all intents and purposes the defeat of Choiseul's campaign against the British fishery at Newfoundland was virtually assured.

Lord Colvill's warships were already in Newfoundland waters when Amherst arrived. Frustrated by his failure to organize a relief expedition in Nova Scotia, Colvill had finally sailed from Halifax on 10 August with *Northumberland* (74), *Gosport* (44) and the Massachusetts provincial armed ship *King George* (20) to join Captain Graves at Placentia. Graves had been waiting there since his arrival in Newfoundland one month earlier. He had been convinced that the French would descend upon Placentia once their conquests elsewhere on the island had been consolidated. With only *Antelope* (50) and *Syren* (20), both of which were old ships, at his disposal, Graves had set the small garrison and his marines feverishly to work improving the harbour defences. Lord Colvill's appearance on 14 August therefore came as a great relief, and Graves looked forward with enthusiasm to abandoning "the defensive to go to seek the Enemy." On 22 August, the warships left Placentia and made for St. John's, where they would cruise in an attempt to sever Ternay's communications with France.[67]

Ternay was not overly disturbed by the appearance of Lord Colvill's little squadron on 24 August. He had been expecting British warships since July and was confident that without troops they could do him no harm. For the next two weeks the British ships were a familiar sight as they cruised before

---

[66]Webster (ed.), *Recapture of St. John's*, 5.

[67]LAC, MG 12, copied from TNA/PRO, ADM 1/482, fols. 214-218, Colville to Clevland, 9 August and 20 September 1762; LAC, MG 13, copied from TNA/PRO, WO 34/42, fols. 97-97v, Colvill to General Amherst, 19 August 1762; and LAC, MG 11, copied from TNA/PRO, CO 194/15, fol. 34v, Graves to Board of Trade, 18 August 1762.

St. John's.[68] His complacency gave way to alarm as Ternay began to receive information that Colvill's warships were about to be joined by an expedition numbering 4500 men. He was certain that St. John's, which he had described as "impenetrable" two months earlier, could not be held against such a force. He therefore convinced d'Haussonville on 8 September to evacuate most of the garrison and prepare to debark immediately for France; 300 men under the command of de Bellecombe would remain to surrender the fort honourably or else hold it through the winter, should the British expedition be delayed or prove to be imaginary. The embarkation was completed by 11 September, but contrary winds prevented Ternay's departure. By the time the wind shifted to a favourable direction the following day, Colonel Amherst's transports had made their appearance, and Ternay realized that the English expedition sent against him was smaller than he had been given to believe. Consequently, he decided to stay at St. John's, after all; the lengthy embarkation was reversed and the grenadiers, together with the marines of the squadron, returned to shore.[69]

Ternay was confident that the British could be repulsed. The narrow entrance to Quidi Vidi Harbour, which was about a mile from St. John's and was the only place where the British could land artillery and stores, had been blocked by sinking several shallops; without artillery, the British could not hope to take Fort William. Ternay also knew that d'Haussonville had stationed detachments of troops at the more distant landing places, so that the British would be unable to achieve the degree of surprise which had made the French capture of St. John's possible. Events, however, were to show that Ternay's confidence was misplaced.[70]

Under the watchful protection of *Syren*, Amherst's transports had made for Torbay, a few miles to the north of St. John's, and anchored there in

---

[68]LAC, MG 2, transcript, 57, copied from AM, B⁴/104, Ternay, "Journal de ce qui s'est passé à St. Jean depuis le 8. Septembre jusqu'au 15 du même mois, Jour de mon départ;" and LAC, MG 12, copied from TNA/PRO, ADM 1/482, fols. 214-218, Colvill to Clevland, 9 August and 20 September 1762.

[69]LAC, MG 2, transcripts, 39-40, copied from AM, B⁴/104, Ternay to Choiseul, 9 July 1762; and Ternay, "Journal," transcript, 57-58.

[70]*Ibid.*, transcripts, 57-59. Other primary accounts of the recapture of St. John's are provided in Webster (ed.), *Recapture of St. John's*; "Col. Amherst's Account," *Gentleman's Magazine*, XXXII (1762), 487-488; and C.H. Little (ed.), *The Recapture of Saint John's, Newfoundland. Dispatches of Rear-Admiral Lord Colvill 1761-1762* (Halifax, 1959). The best secondary accounts of the events of 1762 include Major Evan W.H. Fyers, "The Loss and Recapture of St. John's, Newfoundland in 1762," *Journal of the Society for Army Historical Research*, XI (1932), 179-215; Georges Cerbelaud Salagnac, "La reprise de Terre-Neuve par les Français en 1762," *Revue française d'histoire d'outre-mer*, LXIII (1976), 211-212; and Linÿer de la Barbée, *Chevalier de Ternay*, chap. 8.

the evening of 12 September. The French detachments there were too small to impede either the landing of the troops in the morning or the march south later that day. An attempt to stop the British advance at the river which empties into Quidi Vidi Harbour was equally unsuccessful. By 14 September, the British were clearing that harbour of its obstacles and had even begun to unload light artillery and stores by shallop, despite a bombardment from a small French battery which had been hastily thrown up on Signal Hill.

When the French were forced from their positions by a daring assault under cover of fog the next morning, Ternay's situation became truly desperate. Not only Fort William but also his entire squadron would be at the mercy of the British once they hauled heavy artillery onto the heights they now controlled. As a result, Ternay abandoned all thoughts of resistance and concentrated instead upon escape. In the fog which continued to enshroud St. John's, he prepared to evacuate the troops from Fort William and take his ships through the harbour narrows. When the wind shifted favourably for an immediate departure, Ternay decided that the additional delay needed to embark the troops might only bring another shift of wind, a shift which would leave him trapped in the harbour. Ternay therefore sailed from St. John's, leaving behind almost all of the French land forces together with most of the marines of the squadron.

Lord Colvill, who did not detect Ternay's escape, peevishly regarded it as a "shamefull Flight," adding that "had the French any naval Honour to loose, this Flight of Monsieur Ternay's woud finish it."[71] Captain Hugh Debbieg, the military engineer attached to Amherst's expedition, offering a more realistic assessment of Ternay's predicament, suggested that:

> Monsieur De Ternai...shewed his wisdom in retreating from St. John's Harbour the moment we got Possession of the Hills he knew well enough, there was much less risque in meeting with Lord Colvill's Fleet, than to remain in the Harbour, where the Fate of his Squadron was so certain.[72]

Ternay's departure from St. John's sealed the fate of the French troops which, under the command of Colonel d'Haussonville, continued to hold Fort William in defiance of British demands to surrender. Only when the British managed to set up a mortar battery and began to shell the fort did

---

[71]Ternay, "Journal," transcripts, 62-64; LAC, MG 12, copied from TNA/PRO, ADM 1/482, fol. 217, Colvill to Clevland, 20 September 1762; and LAC, MG 11, copied from TNA/PRO, CO 194/26, fol. 13, Colvill to William Amherst, 16 September 1762.

[72]WLCL, Shelburne Papers, vol. 86, fol. 31, Debbieg to the Board of Ordnance, 8 January 1766.

d'Haussonville agree to capitulate. By the end of 18 September, St. John's was once more in British hands.[73] As for Ternay, his squadron was too under-manned and wracked with disease to attempt any further attack upon British trade or territories. It headed instead for Brest and, when that refuge proved to be too well guarded by British cruisers, for Corunna.[74]

Ternay would remain there for several months until the war dragged to an end, trying to convince Choiseul, and perhaps himself, that his mission had been a success. The cost to the British economy of the interrupted cod fishery alone, he estimated at £1 million. Some 460 vessels of all kinds, to-gether with the shore facilities of the fishery, had been destroyed. Over 100 hostages had been taken, 350 Irishmen had been recruited into French service and most of the residents of St. John's had been expelled. An English frigate had been captured and added to Ternay's squadron, and another ship, an Eng-lish privateer, had been taken just before entering Corunna. "All these losses," he concluded, "cause me to view this campaign as being completely advanta-geous to the State."[75]

To a limited extent, Ternay was correct in insisting that his attack on Newfoundland had been a blow to the fishery. The physical damage to the boats and facilities there would take time and money to repair. The effect on the trade, which had already suffered a setback when Spain entered the war and closed that country as a market for British fish, was equally serious. Al-though the news that St. John's had been recaptured created "universal joy" among the English merchants of the Italian markets, it was highly unlikely that very much of the trade for 1762 could be salvaged.[76] The French occupation had completely disrupted the fishing industry in Newfoundland and frightened off many of the employers and merchants "till it was too late for the fish-ery."[77] Nevertheless, the damage was much less severe than it could have

---

[73]Webster (ed.), *Recapture of St. John's*, 11; and "Col. Amherst's Account," 485.

[74]LAC, MG 2, transcripts, 51-52 and 65-66, copied from AM, B4, vol. 104, Ternay to Choiseul, 5 and 23 October 1762; and Linÿer de la Barbée, *Chevalier de Ternay*, 169-171.

[75]LAC, MG 2, transcripts, 51-56, copied from AM, B⁴/104, Ternay to Choiseul, 5 and 15 October 1762.

[76]Lewis (ed.), *Walpole's Correspondence*, XXII, 5, Mann to Walpole, 6 Feb-ruary 1762; 100n, Dick (Leghorn) to Egremont, 5 November 1762; and 99-100, Mann (Florence) to Walpole, 13 November 1762.

[77]LAC, MG 17, transcripts, 46-50, copied from SPG, B6, Langman to Bur-ton, 2 November 1762; and LAC, MG 11, copied from TNA/PRO, CO 194/15, fols. 42-43v, Graves to Board of Trade, 4 October 1762.

been. The French detachment which had been instructed to destroy the fishery to the south of St. John's evidently stopped short of Ferryland, where British marines belonging to *Syren* had been stationed in June and later reinforced.[78] A significant portion of the British fishery at Newfoundland was therefore spared. Even in those parts of the island where the French did appear, many of the merchants had been able to send away their best goods and effects before they could be confiscated. Moreover, many of the warehouses were left intact, together with their contents, after Ternay decided to hold St. John's. Had they been destroyed according to Ternay's instructions, the injury to the British fishery would have been very severe.

In earlier wars the fishery had survived repeated destruction of its shore facilities because it was then still a migratory fishery with relatively few permanent structures or investments on the island. This was no longer the case by 1762. Extensive storage, service and trading facilities in many of the harbours gave proof that the age of the so-called "great merchant" had arrived. Some of the establishments were worth tens of thousands of pounds. More importantly, the fishery was developing a considerable dependence upon the presence of permanent facilities in Newfoundland. Their destruction would therefore have been a considerable blow to the fishery.[79] Finally, a large number of captured ships and vessels were found in the harbour and retaken after the French were defeated.[80] In purely quantitative terms, then, Ternay had failed to damage the fishery as severely as he had claimed.

---

[78]LAC, MG 12, copied from TNA/PRO, ADM 1/1835, part 3, Graves to Clevland, 18 August 1762.

[79]When Joseph White, the Poole merchant with extensive holdings at Trinity, died in 1771, the estate was valued at £150,000; W. Gordon Hancock, "The Poole Mercantile Community and the Growth of Trinity, 1700-1839," *Newfoundland Quarterly*, LXXX, No. 3 (1985), 22. The presence of a resident fishery helps explain why the British fish trade overshadowed that of France during the eighteenth century; see Christopher Moore, "The Markets for Canadian Cod in the Eighteenth Century: France's Cod Trade and the Problem of Demand" (Unpublished paper presented at the Annual Meeting of the Canadian Historical Association, University of Guelph, 1984), 18-19. The rise of the resident fishery is described in Matthews, "History," 365-368 and 378-381.

[80]LAC, MG 22, copied from TNA/PRO, CO 194/15, fols. 42-43v, Graves to Board of Trade, 4 October 1762. For an example of how one British merchant responded to the French attack, see the survey of Benjamin Lester's career at Trinity in W. Gordon Hancock, "A Biographical Profile of 18th and Early 19th Century Merchant Families and Entrepreneurs in Trinity, Trinity Bay" (Unpublished manuscript at Centre for Newfoundland Studies, Memorial University of Newfoundland, 1980), esp. 91-92.

Responsibility for this failure must rest squarely with Ternay himself. While Choiseul condoned Ternay's decisions even when they went against his instructions, the minister of marine was hardly in a position to influence them as they were made. When Ternay presented him with a *fait accompli*, Choiseul presumably decided to make the best of what limited success Ternay had achieved. If any blame can be placed on the minister, it was for appointing Ternay in the first place. The initiative and courage which Ternay had demonstrated in rescuing *Robuste* and *Eveillé* from the Vilaine were desirable qualities, but so too were experience, a sound grasp of strategical principles and steadiness in the face of an unpredictable situation which, in Ternay, were not proven qualities. His determination to keep his warships out of English hands was admirable; at one point he had told Choiseul that "they didn't leave the Vilaine River just to be captured at St. John's."[81] But such sentiments resulted in cautious tactics and may even have influenced his decision to hold St. John's. Ternay's uncertainty was most evident as he vacillated between resistance and flight in the face of English measures to recover St. John's. Time and again, he behaved as the junior captain he actually was, more accustomed to the relatively limited responsibility of commanding a single warship than as the *chef d'escadre* Choiseul had made him, responsible for the leadership of a combined expedition.

Ternay's weak grasp of strategical principles was, however, his greatest flaw. His mission had been to conduct a series of raids against British trade and possessions in the North Atlantic. The descent on Newfoundland was conceived as an indirect assault on Great Britain's ability to finance the war. As such, Choiseul's plan was indicative of the way in which the French were capable of glorious gambles which hinted at an intuitive understanding of the limitations of British sea power. Yet Ternay's decision to transform his mission into one of conquest shows that he, at least, did not grasp the full implications of those principles of maritime strategy which the Royal Navy was on the way to mastering. Thus, so long as he adhered to Choiseul's plan, Ternay had demonstrated brilliantly that local command of the sea, which the British were exercising so thoroughly by 1762, did not preclude the continuation of French maritime operations. Ternay did not need a squadron capable of disputing command of Newfoundland waters with local British forces because "command," as Corbett would later argue, did not mean the occupation or control of the sea itself but rather of sea communications.

So long as his objectives remained unclear, the British response to Ternay's activities remained a limited one. Only when Ternay departed from his instructions by attempting to establish a French base at St. John's was General Amherst given both a fixed target and the time to prepare a counter-

---

[81]LAC, MG 2, transcripts, 40-41, copied from AM, B⁴/104, Ternay to Choiseul, 9 July 1762.

stroke. Moreover, by remaining in one place Ternay had to abandon the rest of his mission; Île Royale and its fishery, and the coasts of Scotland and Ireland were left untouched. Although the negotiations which brought the war to an end followed close on the heels of the French capture of St. John's, no one, including Ternay (who had the most to gain by advancing such a claim), ever suggested that the one contributed to the other. The moment St. John's became the object rather than an objective of Ternay's expedition, his mission lost any hope for success.

Showing the Flag:
Hugh Palliser in Western Newfoundland, 1763-1766[1]

In the summer of 1764, something rather remarkable occurred in the Bay of Islands on the western coast of Newfoundland. The bay had long been a favourite destination of Breton fishing vessels from Saint-Malo and Basque craft from Saint-Jean-de-Luz and Bayonne. Technically, they should not have been there at all: ever since France signed the Treaty of Utrecht in 1713, its fishermen were prohibited from fishing except on the so-called "French" or "Treaty Shore" to the north. But the fishing grounds in western Newfoundland were rich and productive, and the Bay of Islands had all the ingredients necessary to maintain seasonal shore stations. Best of all, despite its claims of sovereignty, the region was devoid of English patrols. British warships stationed at Newfoundland were virtually unknown north of Bonavista or west beyond Placentia. Western Newfoundland was hardly known to them at all before 1763 when *Lark* under Captain Samuel Thompson appeared and remained on station throughout the summer. For the prescient, *Lark*'s visit was the shape of things to come; late in July 1764, not one but three British warships sailed into the Bay of Islands, shortly to be joined by a fourth. Even more remarkable, the squadron was under the command of Captain Hugh Palliser, commander in chief of the Newfoundland station and civil governor of the island.[2] Never before had a governor ventured more than a couple of days' sail from St. John's. There could be no mistake about the message that Palliser was delivering: England was no longer indifferent to its claim to Newfoundland beyond the Avalon Peninsula. It was a message only the Royal Navy could have delivered.

During the eighteenth century, both France and England held the migratory fishery at Newfoundland in the highest regard. In the conventional wisdom of the age, it was a "nursery for seamen," not only employing thousands of landsmen every year but also transforming them into experienced

[1]This essay appeared originally in *The Northern Mariner/Le Marin du nord* III, No. 3 (July 1993), 3-14. This paper is based in part on research undertaken with the support of the Social Sciences and Humanities Research Council of Canada.

[2]*Guernsey*, *Tweed* and *Pearl* arrived at the Bay of Islands on 24 July 1764, anchoring at York Harbour; *Zephyr* appeared on 29 July but did not anchor until the next day. See Great Britain, National Archives (TNA/PRO), Admiralty Papers (ADM) 51/4210, No. 3, captain's log, *Guernsey*; ADM 51/674, No. 1, captain's log, *Pearl*; ADM 51/1007, No. 4, captain's log, *Tweed*; and ADM 51/1099, No. 1, captain's log, *Zephyr*. See also ADM 52/1391, master's log, *Pearl*.

mariners who were then available for service in their national navies in the event of war. This made it a strategic asset of the first order. In terms of direct employment, consumption of domestic goods and services and the generation of a favourable balance of trade with other mercantile powers, the fishery was a source of great national wealth and, by extension in a mercantile age, of great national power. As a result, the fishery was so highly prized by both countries that neither would willingly give it up, either in whole or in part. During attempts in 1761 to negotiate an end to the Seven Years' War, members of the British and French governments independently ventured the same opinion: that the Newfoundland fishery was more valuable than Canada and Louisiana combined "as a means of wealth and power."[3] Yet this perception ironically committed the British government to a policy that discouraged settlement in Newfoundland. Only as long as they returned to England at the end of each fishing season would the fishermen be available if needed for service in the Royal Navy; only if the fishery's goods, services and trade remained centred on England would the fishery benefit the mother country economically. Newfoundland was therefore perceived as an activity rather than a place. As one imperial administrator later explained, "the Island of Newfoundland has been considered, in all former Times, as a great English Ship moored near the Banks during the Fishing Season, for the Convenience of the English Fishermen."[4]

At least two consequences ensued from this perception. First, it meant that administrative and regulatory measures introduced to Newfoundland were designed first and foremost to serve the needs of the fishery, not the people living on the island. Military garrisons for defence, civil officers to govern the people, magistrates and constables to maintain the peace and recognition of

---

[3]The British Board of Trade maintained that "the Newfoundland Fishery as a means of wealth and power is of more worth than both of the aforementioned provinces;" Gerald S. Graham, "Fisheries and Sea Power," *Canadian Historical Association Annual Report 1941*, reprinted in G.A. Rawlyk (ed.), *Historical Essays on the Atlantic Provinces* (Ottawa, 1967), 8. At approximately the same time, the French Minister of Marine, the duc de Choiseul, insisted that "the codfishery in the Gulf of St. Lawrence is worth infinitely more for the realm of France than Canada or Louisiana;" cited in Max Savelle, *The Origins of American Diplomacy: The International History of Anglo-America, 1492-1783* (New York, 1967), 475n. See also Jean-François Brière, "Pêche et politique à Terre-Neuve au XVIIIe siècle: La France véritable gagnante du traité d'Utrecht?" *Canadian Historical Review*, LXIV, No. 2 (1983), 168-170.

[4]Statement of William Knox, former Under Secretary of State for the American Department, given in evidence before the Committee appointed to inquire into the state of trade to Newfoundland, 1793 (*Second Report*, 16-20), reprinted in Great Britain, Privy Council, *In the Matter of the Boundary Between the Dominion of Canada and the Colony of Newfoundland in the Labrador Peninsula* (London, 1926-1927), IV, joint appendix, part X, No. 722.

property rights: all were introduced reluctantly and with severe restrictions to minimize as much as possible any interference with the migratory fishery and to discourage a settled population. Thus, when British officials authorized the appointment of a governor in 1729, the responsibility was assigned to the commander in chief of the warships stationed at Newfoundland for the protection and supervision of the fishery, and only for the duration of the fishing season. The governor was no more resident than the majority of fishermen.[5] Second, out of recognition that the fishery was as valuable and important to French as to English power, and that it was accordingly a vital element in the maintenance of a balance of power in the North Atlantic, the British government generally took care to limit, but not to exclude entirely, France from the fishery. Consequently, when British sovereignty over Newfoundland was acknowledged in 1713 by the Treaty of Utrecht, France was accorded fishing privileges on a stretch of coast between Cape Bonavista and Pointe Riche (see figure 1).[6] Not even a determined foe of French power like William Pitt the Elder could reverse this position fifty years later when negotiating an end to the Seven Years' War. The duc de Choiseul insisted that continued access to the Newfoundland fishery was a *sine qua non* to peace; Pitt's refusal to bend in the matter would contribute to the emergence within the British cabinet of more moderate voices and the eventual resignation of Pitt as prime minister.[7] Nevertheless, there was a widespread perception that access to the fishery was critical to the ability of France to challenge British power in the North Atlantic. As a result, France was determined after 1763 to push its rights and privileges in the fishery as far as interpretation of the Treaty of Utrecht would permit, while England was equally firm in insisting that those same rights and privileges be observed in as restrictive a fashion as possible.

This had not always been the case. During the decades after 1713, the terms relating to the Treaty Shore had been honoured more in the breach than in the observance. The English fishery and even the settlement frontier had pushed into Notre Dame Bay, which was within the eastern limits of the

---

[5]See Christopher English, "The Development of the Newfoundland Legal System to 1815," *Acadiensis*, XX, No. 1 (1990), 89-119, esp. 97ff.

[6]See Brière, "Pêche et politique," 168-173, and James K. Hiller, "Utrecht Revisited: The Origins of French Fishing Rights in Newfoundland Waters," *Newfoundland Studies*, VII, No. 1 (1991), 23-39. What the Treaty of Utrecht failed to make clear was whether French fishermen had exclusive or concurrent rights to the use of the Treaty Shore. This failure would lie at the heart of what subsequently became known as the "French Shore Question."

[7]James K. Hiller, "The Newfoundland Fisheries Issue in Anglo-French Treaties, 1761-1783" (Unpublished paper presented at the Ninth Atlantic Canada Studies Conference, St. John's, Newfoundland, May 1992).

French Shore. Yet French fishermen were not immediately concerned for they did not then fish in the region. As long as there was no direct contact between English and French fishermen, there was no problem.[8] Similarly, Basque, Breton and to a lesser extent Norman fishermen made full use of the western Newfoundland coast during the decades after 1713 – Laurier Turgeon referred to the region as *"le fief des Basques"* – because the area was virtually ignored by the English.[9] The British government was far more concerned by the slow rate at which English fishermen were extending their activities onto the newly-acquired south coast.[10] Though the west coast was rich in fish, furs, timber and other resources, according to a 1715 survey conducted by William Taverner, he conceded that he had not visited the region himself but had relied instead on a Basque fishing captain for his information. A more characteristic British attitude was that of the military commandant at Placentia who reacted with indifference to reports that French fishing ships were active on the west coast, dismissing the region as a place of "Rocks and Foggs."[11] Thus, the west coast remained virtually a *terra incognita* to English fishermen throughout the first half of the eighteenth century.

---

[8]Brière, "Pêche et politique," 172-173.

[9]Laurier Turgeon, "Les échanges franco-canadiens: Bayonne, les ports basques, et Louisbourg, Ile Royale (1713-1758)" (Unpublished mémoire de maitrise, Université de Pau, 1977), 43; and Laurier Turgeon, "La crise de l'armement morutier basco-bayonnais dans la première moitié du XVIIIe siècle," Société des Sciences Lettres et Arts de Bayonne, *Bulletin*, nouvelle série, No. 139 (1983), 80-85.

[10]To explain the seeming reluctance with which Placentia Bay and the coast beyond the Burin Peninsula were incorporated into the normal domain of the English fishery, Ralph Greenlee Lounsbury blamed "the inherent conservatism of the inhabitants and West Countrymen," as well as the persistence of the former French inhabitants who retained their properties by swearing the oath of allegiance; Lounsbury, *The British Fishery at Newfoundland, 1634-1763* (New Haven, 1934; reprint, New York, 1969), 246. More recently, C. Grant Head has suggested that the abuse of authority by the military establishment at Placentia was to blame. He singles out Colonel Samuel Gledhill, the military commandant and lieutenant governor of Placentia, 1719-1727, whose "interference at Placentia town itself, key place of the whole southern coast, may have been a large contributing factor to the area's slow development under the British." Head, *Eighteenth Century Newfoundland; A Geographer's Perspective* (Ottawa, 1976), 60.

[11]TNA/PRO, Colonial Office Papers (CO) 194/6, 240-241, Captain DeHaldy's information, in William Taverner, "Second Report on Survey, Cape St. Mary's to the South Coast," n.d. (February 1715?, received 20 May 1718); and CO 194/6, 302, Lt. Col. John Moody to Mr. Popple, Secretary to the Board of Trade, 20 August 1719.

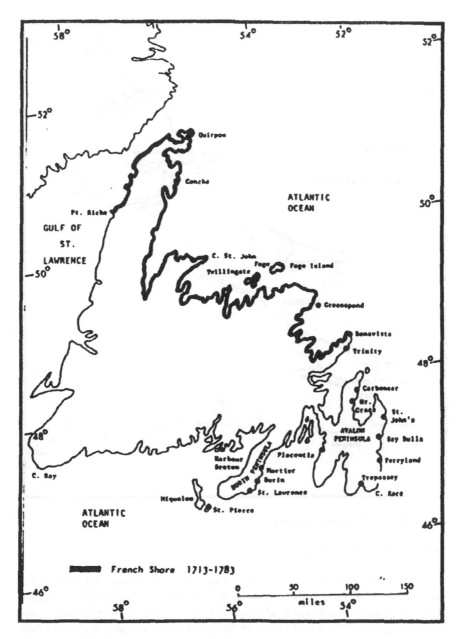

Figure 1:   Eighteenth-Century Newfoundland

Source: Courtesy of the author.

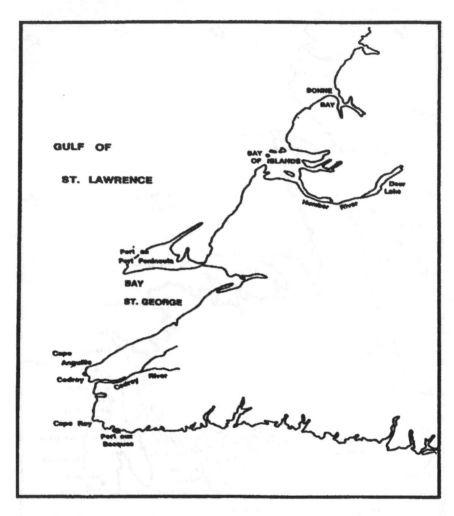

Figure 2:   Southwestern Newfoundland

Source: Courtesy of the author.

One exception to this general rule arose out of the appearance of a permanent community, consisting of about a dozen families of mixed French and Irish origins, which was established at Codroy in southwestern Newfoundland during the 1720s and 1730s (see figure 2). It was maintained by French metropolitan merchants and probably Anglo-American traders.[12] The settle-

_____

[12]Olaf Uwe Janzen, "'*Une petite Republique*' in Southwestern Newfoundland: The Limits of Imperial Authority in a Remote Maritime Environment," in Lewis R.

ments became a matter of some concern to the British government, which instructed Lord Waldegrave, the English ambassador to France, to complain that the Codroy inhabitants represented an infraction of the Treaty of Utrecht.[13] The French government responded with assurances that the people of Codroy had not settled in Newfoundland with the permission or support of French authorities but were *"Brigands déserteurs"* from the French fishery, whose continued presence was an irritant to French efforts to develop the fishery and colony at neighbouring Île Royale.[14] A visit to Codroy by HMS *Roebuck* in 1734 quickly reassured the authorities that English sovereignty over the region was not at risk, for the inhabitants claimed they had sworn an oath of allegiance to the local authorities at Placentia. Moreover, they were described as "Miserably poor, and have neither Fortification or arms, and ...the Coast is so dangerous, that it is not safe for any thing above a Sloop to venture in with the Land."[15] The clinching argument was made when the French offered to expel the Codroy settlers and burn their homes.[16] England was not anxious to allow France to exercise such powers in British territory, and any question that the Codroy settlers might represent a violation of the Treaty of Utrecht was quietly dropped. When war broke out a few years later, the Codroy settlement was abandoned. Protected by neither France nor England and exposed to the maraudings of privateers, the inhabitants fled for the seeming safety of Louisbourg. The community was re-established after 1748, but this time the British were less inclined to tolerate its existence. When the hostilities which would usher in the Seven Years' War began in 1755, warships were sent to southwestern Newfoundland, where they seized fishing vessels, destroyed dwellings, stages, and fishing craft, and removed more than sixty inhabitants, mostly women and children, landing them at Gabarus Bay near Louisbourg.[17]

Fischer and Walter Minchinton (eds.), *People of the Northern Seas* (St. John's, 1992), 1-33.

[13]British Library (BL), Additional Manuscripts (ADD) 32,785, ff. 117-117v, Waldegrave (Paris) to the Duke of Newcastle, 12 June 1734 (N.S.).

[14]BL, ADD 32,785, ff. 170-170v, "Memoire pour servir de Reponse...," 14 June 1734.

[15]TNA/PRO, ADM 1/1498, No. 1, Lord Muskerry (St. John's) to Mr. Burchett, Secretary of the Board of Trade, 19 September 1734.

[16]TNA/PRO, State Papers (SP) 78/207, 67v, memorial enclosed in J. Burnaby (Paris) to John Courand, 2 March 1735 (N.S.).

[17]TNA/PRO ADM 1/481, 68, Admiral E. Boscawen to the Admiralty, 15 November 1755. For details of the operation, see ADM 51/940, ii, captain's log, HMS *Success* (March 1754-May 1757); and ADM 51/3770, iii, captain's log, HMS *Arundel*.

When settlement resumed at Codroy in the early 1760s, it would be sponsored by an English merchant under the Union Jack.

The questions that arose over the settlements at Codroy and Port aux Basques, and the subsequent responses of the British and French governments, would suggest that both England and France agreed over the extent of British sovereignty over western Newfoundland. Certainly the British claimed jurisdiction, an avowal which the French government had tacitly acknowledged in 1734 and 1735. Yet French outfitters continued to send their fishing vessels to the region in the 1730s and 1740s, so that for the fishery at least the matter had not been settled. On those rare occasions when they gave the matter any conscious consideration, they clearly felt that the west coast was a legitimate part of the French Shore, largely because the Treaty of Utrecht had set its western limit at Pointe Riche and no one would admit precisely where that was.[18] Thus, when the merchants of Saint-Malo learned in 1733 that the British might begin enforcing their privileges in the fishery more vigorously, they feared that out of ignorance for the precise limits of the French and British fishing zones their vessels might trespass into forbidden territory and be arrested. They therefore desired that the French government clarify the location of Pointe Riche "*que personne ne connoit sous ce nom à moins que l'on n'entende parler du cap languille proche le cap de rais. Vis à vis de L'Isle du Cap Breton ou Isle royalle. En ce cas, il nous Serait permis de pescher tout Le long de la Coste d'occidentalle de la ditte Isle De terreneuve...*"[19]

Adding to the ambiguity was the lack of consistency or accuracy in French maps and charts depicting western Newfoundland. For various reasons, the French in the mid-eighteenth century were only beginning to produce reli-

---

A French perspective of this operation is provided by Jean Lafreche, captain of *Deux Maries* out of Ciboure; see his statement of 9 October 1755 in Archives de la chambre de commerce de Bayonne (ACCB), I.2/40, "Registres du greffe de l'amirauté etably a Ciboure." The officer in charge of the expulsion of the Codroy settlers was Captain Rous of HMS *Success*, whose next assignment upon returning from Newfoundland would be the more infamous expulsion of the Chignecto Acadians; see W.A.B. Douglas, "John Rous," in *Dictionary of Canadian Biography, III: 1741 to 1770 (DCB)* (Toronto, 1974), 572-574.

[18]William H. Whiteley, *James Cook in Newfoundland, 1762-1767* (St. John's, 1975), 24.

[19]Archives de la Marine, Paris (AM), B³/361, 530-531, memorial from the merchants of Saint-Malo to the Minister of Marine, Comte de Maurepas, 12 April 1733.

able charts of North American waters based on scientific principles.[20] While the cartographic survey of the Gulf of St. Lawrence in 1750 by Joseph de Chabert produced excellent charts, those with the greatest impact were drawn by Joseph-Nicholas Bellin, who was employed for three decades by the French *depot des cartes* to prepare a series of nautical charts that began to appear in atlases in the early 1750s. Bellin's earliest charts do not identify Pointe Riche, though Cape Ray is depicted accurately. Other French charts of this era placed Pointe Riche correctly near Port-aux-Choix or, as it was sometimes called, "P. de Choard."[21] But by 1764, Bellin's charts identified Cape Ray as Pointe Riche, and the French government was using them in an attempt to clarify the ambiguities concerning Pointe Riche in its favour.[22]

French assertion that the west coast lay within its privileged domain was motivated by a very real concern for the future of its sea power. French fishermen returning to Newfoundland after the Seven Years' War discovered that the encroachment of the English on the French shore now extended north beyond Cape St. John and threatened the heart of the area favoured by the French, which they knew as "*le Petit Nord.*" British warships were also assigned to patrolling stations far beyond the usual limits of the British fishery, and for the first time they were burning shallops and cabins left when French fishermen returned home at the end of the season.[23] Petitions to the Minister of

---

[20]James Pritchard, "The Problem of North America in French Nautical Science During the 17th and 18th Centuries," in Martine Acerra, *et al.* (eds.), *Les marines de guerre européennes XVII-XVIIIe siècles* (Paris, 1985), 331-344.

[21]See, for instance, AM, 6JJ/37, No. 66, Joseph-Nicholas Bellin, "Carte de L'Isle de Terre-Neuve," Bellin, 1742. There are several French charts in the cartographic collection of the National Maritime Museum in Greenwich, which also place Pointe Riche near Port-aux-Choix; see, for instance, Vz 9/35, a French chart of 1753.

[22]According to Whiteley, the French ambassador supported his country's claim "on several maps of Newfoundland, chiefly on ones of Herman Moll, published in 1715 and 1720." See William H. Whiteley, "James Cook, Hugh Palliser and the Newfoundland Fishery," *Newfoundland Quarterly*, LXIX, No. 2 (1972), 20. If so, the claim was expressed by more recent and authoritative maps, such as those by Bellin. The copy of his "Carte Reduite du Grand Banc et d'une partie de l'isle de Terre-Neuve" cited in note 21 is especially curious. It was done entirely in black and white, except for the stretch of coast to the north of Cape Ray, which was done in red and marked "*à corriger.*" This part of the original chart had been removed and carefully replaced, presumably signifying the "correction." The result, when compared to Bellin's earlier charts, identifies Cape Ray as Pointe Riche and places Bay St. George much further south than it should be.

[23]Until *Lark* arrived during the summer of 1763, no British warship had ever been stationed there.

Marine, the duc de Choiseul, warned that these actions were part of a deliberate English strategy to prevent France from re-establishing its navy by forcing
fishermen to abandon the fishery and trade.[24] These were words guaranteed to
capture Choiseul's attention, for his willingness in 1762 to continue a losing
war rather than cede France's "nursery for seamen" was a matter of record.[25]
Within a few years, Choiseul would introduce cash incentives or bounties to
outfitters in the fishing industry as part of a programme to rebuild his country's navy. Such direct government encouragement of the fishery was extremely unusual and suggests how closely Choiseul associated the health and
vigour of the French fishery with the navy and national sea power. Choiseul
also recognized that many of the details in the Treaty of Utrecht were poorly
defined and was prepared to use that ambiguity as part of a campaign to secure
the French presence in the fishery at Newfoundland.[26]

   The British government, of course, disagreed that Pointe Riche lay as
far south as Cape Ray and took immediate steps to challenge the French claim.
Moreover, a major redirection of policy was also occurring within the British
government towards its overseas trades and possessions. According to Jack
Greene, the change "amounted to a shift on the part of imperial authorities
from a posture towards the colonies that was essentially permissive to one that
was basically restrictive."[27] Though Newfoundland continued to be perceived
more as a fishery than a settled society, the island's transformation into a colony had already begun, placing pressure on government to interfere more frequently in its affairs. After the Seven Years' War, government involvement in
the fishery became both significant and persistent. The most visible expression
of this shift was the decision to appoint Hugh Palliser as governor and commander in chief of Newfoundland, beginning in 1764. Palliser would serve
longer at Newfoundland than any other eighteenth-century governor and was

---

[24]"*C'est un parti-pris en Angleterre de faire la pesche et la sécherie des
morues sur cette coste concurremment avec les François, afin de les forcer successive-
ment et par des pertes suivies à abandonner entierrement ce commerce, dont ils sentent
bien toutte l'influence sur le rétablissement de la marine françoise...le but [de la
politique anglaise] est d'empescher le rétablissement de notre marine...*," cited in
Brière, "Pêche et politique," 174. See also Hiller, "Utrecht Revisited," 34-35.

[25]See Hiller, "Newfoundland Fisheries Issue," 3-10.

[26]In his seminal study of *The French Shore Problem in Newfoundland: An Imperial Study* (Toronto, 1961), Frederic F. Thompson suggests that the attempt to identify Cape Ray as Pointe Riche was a negotiating ruse of 1783 (15). In fact, it was a
pressure tactic applied in 1763-1764.

[27]Jack P. Greene, "An Uneasy Connection; An Analysis of the Preconditions
of the American Revolution," in Stephen G. Kurtz and James H. Hutson (eds.), *Essays
on the American Revolution* (Chapel Hill, NC, 1973), 72.

far more energetic than any of his predecessors in exercising his authority and carrying out his responsibilities in the fishery.[28] With a determined administration in London and an equally resolute governor, the assertion of British sovereignty within the fishery was predictable.

Palliser began his task almost immediately. Months before sailing for Newfoundland, he sought proof for the British interpretation of the geographical limits of the French Shore in London map shops. He pressed James Cook into this service, for he was well acquainted both with the Yorkshireman's cartographic skills and his familiarity with Newfoundland.[29] Together, the two managed to compile a list of maps and atlases which satisfied them that French claims concerning the coincidence of Pointe Riche and Cape Ray were unfounded; as one historian later remarked, "Palliser was triumphant in rebutting the claim of the French ambassador that Cape Ray and not Point Riche was the really intended southern limit in the west coast of French operations."[30] The British Board of Trade quickly submitted a report on the geographic limits of the French Shore based on Palliser's findings. But they also included in their report the claim that "many of Your Majesty's Subjects have long been used to frequent and carry on a fishery in those Harbours, particularly that near Cape

---

[28]See William H. Whiteley, "Governor Hugh Palliser and the Newfoundland and Labrador Fishery, 1764-1768," *Canadian Historical Review*, L, No. 2 (1969), 141-163; see also Whiteley's essay on Palliser in *DCB, IV* (Toronto, 1979), 597-601.

[29]Cook had been master's mate of Palliser's ship *Eagle* from 1755 to 1757. Later, as master of *Northumberland*, Cook's path crossed Palliser's again when both participated in the successful British effort to recapture St. John's from the French; Glyndwr Williams, "James Cook," in *DCB, IV*, 162-167; and Whiteley, "Hugh Palliser," 597-601. Cook used the opportunity to survey St. John's harbour. That was not unusual; because the eighteenth-century Royal Navy did not provide its warships with charts, it was a poor sailing master who did not make his own whenever he visited a new coast or harbour. What was unusual was the impressive quality of Cook's charts, and he was soon brought to the attention of Governor Thomas Graves. As Cook would later acknowledge, it was Graves who secured Cook's employment in surveying and charting the Newfoundland coast, beginning in 1763 with the south coast of Newfoundland and the French islands of St. Pierre and Miquelon, and continuing until 1767 with the west coast; National Maritime Museum, Greenwich (NMM), Graves Papers (GRV), vol. 106, unpaginated, James Cook to Thomas Graves, 15 March 1764: "It is more than probable [that] the Survey of the Island will go on untill compleatly finished, this usefull and necessary thing the World must be obliged to you for." See also J.C. Beaglehole, *The Life of Captain James Cook* (Stanford, 1974), esp. 62-66.

[30]Beaglehole, *Life*, 77. See TNA/PRO ADM 1/2300, No. 9, Captain Palliser's Letters, Cook to Palliser, 7 March 1764, and "A List of Maps and Charts in which Cape Ray or Pointe Riche or both have been inserted," 15 March 1764; Beaglehole reprints Cook's letter in its entirety.

Ray and in the Bay of St. George."[31] This was stretching the truth, for the only "Subjects" of King George III in the region were connected with the settlement at Codroy, and as even the Board of Trade should have known, that community had been French until 1755.[32] It was to add weight to these conclusions that Hugh Palliser was directed to "visit all the Coasts and Harbours of the said Islands and Territories under Your Governmt," a directive which took Palliser and most of the ships in his command to the Bay of Islands during the summer of 1764.[33]

Palliser's little squadron departed St. John's on 4 July and made its way to the west coast by way of Placentia and St. Pierre, a small island lying off the south coast of Newfoundland and which, together with the adjacent island of Miquelon, had been turned over to France in 1763 to serve its fishermen as an *abri* or shelter. Palliser had been disturbed by reports that several terms of the treaty which restored those islands to France were being violated.[34] An exchange of correspondence with the French governor had ensued which gave Palliser the opportunity "to discribe the true Situation of Pointe Riche to be in the Latitude of 50°, 30' N., and gave them to understand that [he] intended to visit those Parts..., by which [he hoped] that Point is now Established to His Majesty's Satisfaction."[35] Yet when Palliser arrived at St. Pierre, he discovered a French squadron under Commodore Tronjoly preparing to sail for the Gulf of St. Lawrence, then north to visit the harbours of the

---

[31]TNA/PRO, CO 195/9, 330-356, Board of Trade to the King, 20 April 1764.

[32]After two months spent patrolling the coast between Port aux Basques and Port-aux-Choix, *Lark* returned to England where Samuel Thompson submitted his report. He indicated that the community was English but mentioned that it had belonged to the French until "the beginning of the late Warr." *Lark*'s movements can be traced using the master's log in TNA/PRO, ADM 52/1316, ii. These indicate that for a few days, at least, the ship was joined by *Tweed* (Captain Perceval). Thompson's report to the Admiralty secretary, Philip Stephens, dated 12 March 1764, appears in TNA/PRO, ADM 1/2590, No. 4, Captain Thompson's Letters. The most detailed description of the Codroy settlement appears in his "Remarks" in NMM, GRV/105, "Answers to Heads of Inquiry (1763)."

[33]TNA/PRO, CO 195/9, 286-287, Board of Trade, draft of instructions for Hugh Palliser, No. 12, 10 April 1764.

[34]In particular, Palliser was concerned by reports of French warships in the area, fortification of St. Pierre and Miquelon contrary to the treaties, trade with New Englanders and French visits to Newfoundland's south coast for wood; see Whiteley, "Governor Hugh Palliser and the Newfoundland and Labrador Fishery," 144-145.

[35]TNA/PRO, SP 42/65, No. 30(b), fol. 384-385, Palliser to Philip Stephens, 1 September 1764.

French Shore. Tronjoly agreed to cancel his plans in the face of Palliser's strong objections, but the incident must have made Palliser more determined to assert England's sovereignty in regions such as the west coast where its presence had heretofore been tenuous at best.

Figure 3:   The Bay of Islands

Source;   Courtesy of the author.

Significantly, Palliser's ships did not stop at Codroy, which was the only settlement of significance on that coast, except to hoist their colours and fire a gun to attract a pilot.[36] Instead, they made directly for the Bay of Islands

---

[36]See, for instance, Admiralty Hydrography Library, Taunton (AHL), Miscellaneous Papers 27(T), Remark Book, HMS *Tweed*, Captain Perceval, Mr. James Maxwell, Master, 24 July-1 August 1764.

(see figure 3). This wide bay had long been a preferred destination for French fishermen, who knew it as *Baie des trois îles,* after the three large islands that stood like sentinels across its entrance. They were attracted by the many smaller coves and harbours, like Little Port and Bottle Cove just outside the bay, and York Harbour and Lark Harbour within the bay. Here was every requirement for a productive shore fishery: locations convenient to excellent fishing grounds; huge schools of capelin and herring to draw the cod inshore in the spring and to provide bait; good stone beaches; wood and water in abundance; warm, sunny summers with steady breezes to dry cod; and perfect shelter from any storms from the Gulf of St. Lawrence. If there were Frenchmen on the west coast in violation of the treaties, the Bay of Islands was where they were most likely to be found. On 24 July, *Guernsey, Pearl* and *Tweed* made their way past Lark Harbour where Captain Thompson had moored his ship the year before and dropped anchor in what became known as York Harbour. A few days later, they were joined by *Zephyr.*

The crews immediately fell to work at the normal routine of a warship on remote station: collecting wood and water; brewing spruce beer; transferring bread from ship to ship; and enjoying the rare luxury of supplementing their provisions with cod and trout found in abundance.[37] Joseph Gilbert, *Guernsey*'s master, busied himself making charts of the bay.[38] This in itself was not unusual, but the generosity with which his charts sprinkled the local landscape with names that were not only English but unmistakably associated with the visiting warships suggests that Palliser had a hand in the business. Perhaps Palliser was inspired by example; the name "Lark Harbour" pre-dates his visit, and he may have assumed that it was a legacy of Thompson's call the previous year.[39] Certainly Palliser can be forgiven the conceit of leaving a souvenir of his squadron's visit in the local toponomy. After his cartographic inquiries in London earlier that year, he must have appreciated the power that names on a map had in defining sovereignty. As a result, the three islands at the entrance to the bay were named Guernsey, Pearl and Tweed; the large, flat

---

[37]AHL Misc. Papers 27(T), Remark Book, HMS *Tweed* at York Harbour, Bay of Islands, 24 July-1 August 1764; various logs, including *Guernsey* (TNA/PRO, ADM 51/4210, No. 3), *Tweed* (TNA/PRO, ADM 51/1007, No. 4), and *Zephyr* (TNA/PRO, ADM 51/1099, No. 1).

[38]See BL, ADD 17,693a, "A Plan of the Bay of Three Islands in Newfoundland...;" and BL ADD 17,693c, "A Plan of York Harbour and Lark Bay, within the Bay of Three Islands in Newfoundland...taken on board HMS Guernsey, June 1764."

[39]The name "Lark Harbour" may in fact pre-date *Lark*'s visit; when the ship arrived at the Bay of Islands for the first time, the master's log reported that he anchored "In a Small harbour in the Bay of Isles cal'd lark harbour." TNA/PRO, ADM 52/1316, ii, 13 July 1763.

island in York Harbour became known as Governor's Island; the powerful river and fjord that debouched into the Bay of Islands were named after the Humber River in England. Two other bodies of water were named after the Rivers Medway and Thames, also in England. Neither was accurate, since both were fjords, not rivers, and perhaps for this reason the names did not stick; all the others survive to this day.[40]

While Gilbert was attaching a permanent signature of an English presence to the Bay of Islands, Palliser searched for evidence that the French had established themselves contrary to the Treaty. The evidence was not hard to find. Palliser reported dwellings and presumably other shore facilities, and "an infinite number of their Traps for taking Furrs," positive proof that the French had overwintered. Local fishermen from Codroy added that the French had not only fished for cod but also for salmon, an activity that would have required extensive shore facilities and wintering crews.[41] English fishermen had evidently begun to follow suit, for they claimed that during the winter of 1762 they had remained in the Bay of Islands and followed the Humber River as far as a great lake, which stretched off into the distance as far as they could see; this would have been Deer Lake. Normally, Palliser did not condone permanent inhabitancy in Newfoundland, believing it injurious to the migratory fishery. In this instance, however, he uttered not a word of criticism; permanent Englishmen were evidently preferred to permanent Frenchmen, and might even discourage the latter from returning.

On 5 August, *Zephyr* unmoored and headed north to Pointe Riche for one last reminder to the French where the western limits of their presence in Newfoundland lay. *Guernsey*, *Tweed* and *Pearl* had already worked their way out of the Bay of Islands a few days earlier. While *Pearl* headed into the Gulf of St. Lawrence, *Guernsey* and *Tweed* made for St. John's. Though Palliser

---

[40]Traditionally, James Cook is given credit for assigning all these English names when he conducted his cartographic survey in 1767; see, for instance, Whiteley, "James Cook in Newfoundland," 21-22. The mistake is obvious when Gilbert's cruder but older maps, with the names already clearly in place, are examined. The "River Thames" has since become Middle Arm, while the River Medway is known to us today as North Arm. Logic does not always dictate toponomy; Gilbert's appropriately named "Harbour Island" has since become "Woods Island."

[41]TNA/PRO, SP 42/65, No. 30(b), fol. 384-385, Palliser to Stephens, 1 September 1764. This testimony may have been gleaned from his pilots, but more likely came from the Codroy settlers who, like the French, were drawn to the Bay of Islands by its wealth of marine resources. According to Captain Thompson's remarks (NMM, GRV/105), "The Isle of Codroy...is settled by upwards of an hundred People, who carry a Fishery on at the Bay of Islands." Palliser's remarks on Joseph Gilbert's completed chart also suggest that additional testimony came from "the Reports of the Cape Breton Indians who the last Winter was at Caderoy..." See BL, ADD 17,693a, "A Plan of the Bay of Three Islands."

would continue to serve as governor of Newfoundland longer than anyone else in the eighteenth century, this visit to the west coast was both his first and last. In 1765 and again in 1767, he devoted attention to Newfoundland's south coast, where illicit contact between the French at St. Pierre, the fishermen of the south coast and Anglo-American traders had become a problem. Similarly, he would visit the north coast and Labrador, where efforts to invigorate the migratory fishery were threatened by abuses by French, Anglo-American and Newfoundland fishermen and traders. In 1766 and 1768, Palliser gave his attention and energies to the fishery's traditional heartland, the Avalon Peninsula.[42] Yet the west coast was not allowed to fall victim again to indifference and neglect. Even though the Newfoundland station could ill afford to waste resources, a warship was regularly assigned to patrol the west coast after 1764.[43] In 1767, James Cook completed his famous cartographic survey of various parts of Newfoundland with a season on the west coast, including a magnificent chart of the Bay of Islands.[44] We can therefore assume that Palliser was satisfied that England's point had been made; documentary and cartographic evidence, supported by a show of force and a toponomic legacy, had confirmed British possession of the west coast.

Certainly the French did not persist in their pretensions to include western Newfoundland within the boundaries of the Treaty Shore. This is not to say that they had given up hope of extending those bounds as far south as Cape Ray. That particular goal, however, would now be assigned to their diplomats who, while conceding that the Treaty Shore did not include the coast south of Pointe Riche, argued that it should be extended south of that point. In exchange for a new definition of the western limits of the Treaty Shore, France suggested that the eastern limit be moved from Cape Bonavista to Cape St. Jean. Although these negotiations were interrupted by the American Revolu-

---

[42]Whiteley, "Governor Hugh Palliser and the Newfoundland and Labrador Fishery," 147-152.

[43]Usually assigned to the Newfoundland station, in addition to the commodore's flagship, were two frigates, two sloops of war and a couple of schooners, brigs or cutters.

[44]As a result of his work in 1762 with the expedition that recovered St. John's from the French, Cook was commissioned by the Admiralty in 1763 to chart several of the coasts of Newfoundland. The work took him five years, beginning with St. Pierre and Miquelon, Newfoundland's Northern Peninsula, the south coast and finally the west coast. Reflecting the political importance of this assignment, Cook's charts were rich with information about existing fisheries and potential new ones. It is perhaps because of the meticulous, even perfectionist nature of this work that Cook is usually given credit – occasionally erroneously, as this paper maintains – for many of the place-names in western Newfoundland. For a full treatment, see previously cited essays by Whiteley together with Williams' essay in *DCB, IV*.

tion, they would form the basis for the re-definition of the Treaty Shore that was incorporated into the Treaty of Versailles in 1783.[45] Consequently, never again in that century would the Royal Navy need to make its presence as strongly felt on the west coast of Newfoundland as it did in the summer of 1764.

From the perspective of Newfoundland history, it is perhaps too easy to dwell upon the unprecedented nature of Palliser's visit to the Bay of Islands. Never before had the governor and commander in chief of the Newfoundland station visited this part of his jurisdiction. Yet, as was recently observed, in peace as well as war the Royal Navy was "the essential instrument of coercion," and the attainment of foreign policy objectives, including the assertion of sovereignty, was one of its principal functions in the eighteenth century.[46] In Newfoundland, possibly more than in any other station, the Royal Navy had long been called upon to play many roles: as an instrument of defence, fisheries administration, judicial and legal administration and social services. For it to function as an agent of diplomacy was therefore quite consistent, both with its performance elsewhere and with the diversity of its role in Newfoundland.

---

[45]Brière, "Pêche et politique," 178-185; and Hiller, "Newfoundland Fisheries Issue," 11-20.

[46]Nicholas Tracy, *Navies, Deterrence and American Independence: Britain and Seapower in the 1760s and 1770s* (Vancouver, 1988), introduction; see also the several essays in Jeremy Black and Philip Woodfine (eds.), *The British Navy and the Use of Naval Power in the Eighteenth Century* (Leicester, 1989).

The Royal Navy and the Interdiction of Aboriginal Migration to Newfoundland, 1763-1766[1]

Abstract

As a result of the Seven Years' War, France lost most of its territorial empire in North America, including Cape Breton Island. At the same time, France reacquired possession of the tiny islands of St. Pierre and Miquelon off the south coast of Newfoundland. This acquisition was intended to provide France with a toe-hold in the North American fisheries, but for the Mi'kmaq Indians, France's former aboriginal allies in Cape Breton Island, the islands of St. Pierre and Miquelon became the only means by which they could maintain contact with the French. During the period 1763-1766, a significant number of Mi'kmaq therefore attempted to move into the Bay d'Espoir-Hermitage Bay area of Newfoundland's south coast, in close proximity to the French islands. This, however, was something that for various reasons neither the French nor the British desired, and so attempts were made to discourage the migration. Ships of the Royal Navy stationed in Newfoundland played a key part in patrolling the area, reporting on the movements of the aboriginals and attempting to enforce British policy restricting their presence in the region. This paper examines these efforts both as an expression of the Royal Navy's peacetime role as a projector of British power and as an agent in the process that eventually saw the aboriginals abandon their attempts to move to Newfoundland's south coast, settling instead in western Newfoundland. The paper is based on research into the activities and procedures of the Royal Navy in Newfoundland after 1763, research that has already led to a number of papers and publications over the years.[2] The main source for this paper will be the documents in

---

[1]This essay appeared originally in *International Journal of Naval History* VII, No. 2 (August 2008) [e-journal: < http://www.ijnhonline.org/issues/volume-7-2008/aug-2008-vol-7-issue-2/ >].

[2]"Hugh Palliser, the Royal Navy and the Projection of British Power in Newfoundland Waters, 1764-68" (Unpublished paper presented at the XIXth International Congress of Historical Sciences, meeting of the International Commission of Maritime History, Oslo, Norway, August 2000); "Micmac Migration to Western Newfoundland," *Canadian Journal of Native Studies*, X, No. 1 (1990), 71-94 (with Dennis A. Bartels); and "Showing the Flag: Hugh Palliser in Western Newfoundland, 1764," *The Northern Mariner/Le Marin du nord*, III, No. 3 (1993), 3-14. An earlier draft of this paper was presented at the Fourth International Congress of Maritime History in Corfu,

the Colonial Office 194 series and the Admiralty papers, held by the National Archives (TNA/PRO) in England.

Introduction

Although the eighteenth-century Royal Navy is best known for its activities in time of war, there has been growing interest in the navy's activities during the several periods of peace which interrupted the wars of that era.[3] As the late David Syrett explained, those activities were "for the most part...a constabulary role to aid and support British foreign policy and overseas trade."[4] This paper will focus upon the efforts of Royal Navy warships stationed in Newfoundland during the 1760s to carry out a number of responsibilities on the island's south coast, including the unanticipated task of interdicting the movement of aboriginals from Cape Breton Island who were attempting to restore contact with the French in St. Pierre, a tiny island off the tip of Newfoundland's Burin Peninsula. For reasons that will be explained, that movement conflicted with both French and British priorities in the region, hence the attempt to discourage the migration and to enforce British policy restricting Mi'kmaq presence in the region. This paper examines these efforts both as an expression of the Royal Navy's peacetime role as a projector of British power and as an agent in the process that eventually saw the aboriginals abandon their attempts to move closer to St. Pierre and to settle instead on the more remote coast of western Newfoundland.

The Context

The events that unfolded on Newfoundland's south coast between 1763 and 1766 were driven by the particular priorities and needs of three distinct parties; before proceeding with an analysis of actual events, it is therefore necessary to explain the context to those events in terms of those parties.

Greece, in June 2004 under the title "The Navy and the Natives: The Royal Navy and the Attempt to Interdict Aboriginal Migration to Newfoundland, 1763-1766."

[3]See, for instance, the essays in Jeremy Black and Philip Woodfine (eds.), *The British Navy and the Use of Naval Power in the Eighteenth Century* (Leicester, 1989), particularly Julian Gwyn, "The Royal Navy in North America, 1712-1776," 129-147.

[4]David Syrett, "A Study of Peacetime Operations: The Royal Navy in the Mediterranean, 1752-5," *Mariner's Mirror*, XC, No. 1 (2004), 42-50. Syrett was looking at the squadron commanded by Commodore Hon. George Edgcumbe in the Mediterranean during the brief peace between the War of the Austrian Succession and the Seven Years' War.

First, there were the French. The Seven Years' War which came to a conclusion in 1763 with the Treaty of Paris had decimated the French overseas empire in North America. Canada, Île Royale (with the fortified town of Louisbourg), the Gulf of St. Lawrence and the Labrador coast were all gone; the only territory left to France was the tiny islands of St. Pierre and Miquelon off the tip of Newfoundland's Burin Peninsula.[5] British territory since 1714, they were now restored to France to support its offshore bank fishery with an *abri* or shelter. The islands would also enable a small residential fishery to revive. As well, France retained the privileges, first defined in the Treaty of Utrecht in 1713, by which French crews were able to maintain a seasonal sedentary fishery in Newfoundland on the so-called "Treaty" or "French Shore," extending from Cape Bonavista to Point Riche.[6] In short, and notwithstanding the enormous loss of empire in North America, France had managed to preserve access to the North American fishery.

This was an extremely significant achievement. France, like Great Britain, prized the fishery as an important economic and strategic asset that contributed to the wealth and power of the state.[7] In negotiating an end to the Seven Years' War, the French Minister of Marine, the duc de Choiseul, had insisted that continued access to the Newfoundland fishery was a *sine qua non*.[8] His decision in 1762 to continue a war that France had clearly lost, by

---

[5]For St. Pierre and Miquelon, see Jean-Yves Ribault, "La pêche et le commerce de la morue aux Îles Saint-Pierre-et-Miquelon de 1763 f 1793," in *Congrès National des Sociétés Savantes, Actes du 91e congrès Rennes 1966* (2 vols., Paris, 1969), I, 251-292; and Ribault, "La population des Îles Saint-Pierre-et-Miquelon de 1763 à 1793," *Revue française d'histoire d'outre-mer*, LIII (1966), 5-66. The Treaty of Paris also saw France turn Louisiana over to Spain.

[6]James K. Hiller, "Utrecht Revisited: The Origins of French Fishing Rights in Newfoundland Waters," *Newfoundland Studies*, VII, No. 1 (1991), 23-39.

[7]The British Board of Trade maintained that "the Newfoundland Fishery as a means of wealth and power is of more worth than both of the aforementioned provinces...;" Gerald S. Graham, "Fisheries and Sea Power," Canadian Historical Association *Annual Report 1941*, reprinted in G.A. Rawlyk (ed.), *Historical Essays on the Atlantic Provinces* (Ottawa, 1967), 8. At almost the same time, the French Minister of Marine, the duc de Choiseul, insisted that "the codfishery in the Gulf of St. Lawrence is worth infinitely more for the realm of France than Canada or Louisiana." Cited in Max Savelle, *The Origins of American Diplomacy: The International History of Anglo-America, 1492-1783* (New York, 1967), 475n. See also Jean-François Brière, "Pêche et politique à Terre-Neuve au XVIIIe siècle; la France véritable gagnante du traité d'Utrecht?" *Canadian Historical Review*, LXIV, No. 2 (1983), 168-170.

[8]The refusal of William Pitt to bend in the matter contributed to the emergence within the British cabinet of more moderate voices and the eventual resignation of Pitt as prime minister; Jonathan R. Dull, *The French Navy and the Seven Years' War*

sending the Chevalier de Ternay on a raiding expedition into the North Atlantic rather than surrender French access to the North Atlantic fishery, gave proof of the fishery's perceived importance to the state.[9] Choiseul's determination after 1763 to rebuild the shattered French navy placed an additional premium on the preservation of the fishery in North America – the fishery was widely assumed to be a "nursery for seamen" – and it was therefore essential that every step and measure be taken to avoid jeopardizing the well-being of the migratory and sedentary fisheries in the northwest Atlantic. On the French Shore, this meant that the French became extremely protective and jealous of their right to fish within the territorial limits defined by the Treaty of Utrecht, with the result that French-English friction, never a problem in the past, became quite serious after 1763.[10] For François-Gabriel d'Angeac, the first governor of the newly restored islands of St. Pierre and Miquelon, this meant treading an exceptionally sensitive line between protecting French imperial interests and complying with British interpretations of French rights and privileges in the fishery.[11]

For the British, the situation after 1763 in North America generally, and within the fishery in particular, posed significant challenges. The territories which France had lost were British gains, with all the administrative, legal, jurisdictional and regulatory headaches that would entail. Moreover, by 1763, a major redirection of policy had begun to occur towards British overseas trades and possessions. According to Jack Greene, the change "amounted to a shift on the part of imperial authorities from a posture towards the colonies that was essentially permissive to one that was basically restrictive."[12]

(Lincoln, NB, 2005), chap. 8; and James K. Hiller, "The Newfoundland Fisheries Issue in Anglo-French Treaties, 1713-1904," *Journal of Imperial and Commonwealth History*, XXIV, No. 1 (1996), 1-23.

[9]Olaf Janzen, "The French Raid Upon the Newfoundland Fishery in 1762 – A Study in the Nature and Limits of Eighteenth-Century Sea Power," in William B. Cogar (ed.), *Naval History: The Seventh Symposium of the U.S. Naval Academy* (Wilmington, DE, 1988), 35-54.

[10]Jean-François Brière suggests that this friction was deliberate, encouraged by the French in hopes of getting the British to re-define the limits of the French Shore; see Brière, "Pêche et politique à Terre-Neuve."

[11]A detailed account of d'Angeac's efforts is provided in Frederick J. Thorpe, "The Debating Talents of the First Governor of Saint-Pierre and Miquelon, François-Gabriel d'Angeac, 1764-1769," *Newfoundland Studies*, XVIII, No. 1 (2002), 61-83.

[12]Jack P. Greene, "An Uneasy Connection; An Analysis of the Preconditions of the American Revolution," in Stephen G. Kurtz and James H. Hutson (eds.), *Essays on the American Revolution* (Chapel Hill, NC, 1973), 72.

Newfoundland at the time was perceived more as a fishery than as a develop-ing colonial society, yet by the middle of the eighteenth century, the island's transformation into a colony had clearly begun to manifest itself. Government responded to these developments by playing a more determined role in affairs at Newfoundland, a trend that became both significant and persistent after the Seven Years' War.[13]

One expression of this trend was the increase in the number and re-sponsibilities of warships stationed in Newfoundland, from an average of two or three per year during the 1720s and 1730s to as many as eight or nine in the 1760s. Where Newfoundland station ships were once content to remain moored in St. John's or Placentia harbours for much of the season, leaving patrol work to hired boats and vessels, warships after 1763 were increasingly stationed in the remote parts of the fishery, including the island's west and south coasts as well as Labrador. Ships even began to over-winter in New-foundland for the first time.[14]

A second expression of greater British determination to protect its interests in the fishery after 1763 was the cartographic work undertaken by James Cook between 1763 and 1767.[15] The recent war and the events that con-tributed to the outbreak of hostilities in North America as early as 1754 had given British authorities a heightened appreciation for the importance of clarity in defining with precision the boundaries and boundary markers between the British and French empires in North America. Cook's cartography was an ex-ercise in asserting sovereignty over stretches of coast with which the British were not yet very familiar.

A third expression of British efforts to assert their interests in New-foundland was the appointment of Hugh Palliser as governor and commander

---

[13]For a thorough treatment of the evolution of government policy and admini-stration in Newfoundland, see Jerry Bannister, *The Rule of the Admirals: Law, Custom, and Naval Government in Newfoundland, 1699-1832* (Toronto, 2003). For Newfound-land's emergence as a settled society, see Stephen J. Hornsby, *British Atlantic, Ameri-can Frontier: Spaces of Power in Early Modern British America* (Hanover, NH, 2005), chap. 3, "Atlantic Staple Regions: Newfoundland, the West Indies, and Hudson Bay."

[14]Janzen, "Hugh Palliser, the Royal Navy and the Projection of British Power;" see also Janzen, "Showing the Flag."

[15]The best overall study of Cook remains J.C. Beaglehole, *The Life of Cap-tain James Cook* (Stanford, 1974). But for Cook in Newfoundland, see William H. Whiteley, *James Cook in Newfoundland, 1762-1767* (St. John's, 1975); Whiteley, "James Cook and British Policy in the Newfoundland Fisheries 1763-7," *Canadian Historical Review*, LIV, No. 3 (1973), 245-272; and Whiteley, "James Cook, Hugh Palliser, and the Newfoundland Fisheries," *Newfoundland Quarterly*, LXIX, No. 2 (1972), 17-22.

in chief of Newfoundland in 1764.[16] Palliser would be vigorous in the execution of his civil and naval authority and responsibilities. As he explained to Admiralty Secretary Philip Stephens, these responsibilities included the preservation of order within the fishery, the suppression of illicit trade in the region between Anglo-American colonials and the French, and finally "To keep the French within the Limits prescribed by Treaties, and thereby prevent them rivaling us in ye valuable Fish Trade, and from raising so great a Marine from the Fishery as in late times they did."[17] His interpretation of French rights and privileges in the fishery was therefore particularly strict.

Caught in the middle of French and British determination to protect and preserve their respective imperial interests in the fishery was a group of people who were just as determined to look after their own interests. For 150 years, the Mi'kmaq Indians had been friends, trading partners and allies of the French in Acadia. That relationship persisted even after much of Acadia fell under British jurisdiction following the seizure of Port Royal in 1710 and the Treaty of Utrecht in 1713. The Mi'kmaq remained loyal allies of the French as late as the Seven Years' War. Only with the French defeat in 1758 did the Mi'kmaq sign treaties with the British, ending centuries of hostility.

The Mi'kmaq were a maritime-adapted people with a seafaring capability sufficient to extend their territorial range as far as the Magdalen Islands in the Gulf of St. Lawrence. It is moot whether that range included a sustained presence in Newfoundland or whether the Mi'kmaq came to Newfoundland only sporadically until they acquired European shallops. What is clear is that by the eighteenth century, the Mi'kmaq were hunting and trapping in southern and southwestern Newfoundland fairly frequently and that, from 1763 on, their presence in Newfoundland had become both substantial and persistent.[18]

---

[16]See William H. Whiteley, "Governor Hugh Palliser and the Newfoundland and Labrador Fishery, 1764-1768," *Canadian Historical Review*, L, No. 2 (1969), 141-163; see also Whiteley's essay on Palliser in the *Dictionary of Canadian Biography, IV* (Toronto, 1979), 597-601 (*DCB*).

[17]Great Britain, National Archives (TNA/PRO), Admiralty (ADM) 1/470, 13v, Palliser (St. John's) to Admiralty Secretary Philip Stephens, 18 October 1767. Palliser was true to his word. In contrast to his predecessors, he personally visited the more remote coasts in his jurisdiction – with most of his squadron to the west coast one year, to the south coast in the vicinity of the French islands in 1765, as well as the northernmost part of the island, where the French retained fishing privileges by treaty, as well as Labrador.

[18]Charles A. Martijn refers to the Gulf of St. Lawrence as a Mi'kmaq "domain of islands;" see "An Eastern Micmac Domain of Islands," in William Cowan (ed.), *Actes du vingtième congrès des Algonquinistes* (Ottawa, 1989), 208-231. See also his "Les Micmacs aux îles de la Madeleine: visions fugitives et glanures ethnohistoriques," and "Voyages des Micmacs dans la vallée du Saint-Laurent, sur la Côte-Nord et

The reasons for this had less to do with the Mi'kmaq subsistence economy than with cultural concerns emerging out of their traditional relationships with the French. The Mi'kmaq who migrated to southwestern Newfoundland from Cape Breton Island had a strong association with Mirliguèche, the site of a mission established by the French in 1724.[19] Sustained contact with the French had made the Mi'kmaq increasingly dependent upon French arms, ammunition, cloth and food presented in ceremonial gift exchanges. These gifts had their origins as "a matter of protocol to cement alliances and trade agreements" but gradually evolved into essential means of Mi'kmaq subsistence.[20] Such dependency worked both ways, for the French came to rely upon the gifts as a relatively inexpensive means of maintaining their influence in the region. The cost of the gifts was certainly cheaper than such alternatives as encouraging immigration and settlement, or greatly increasing the French military establishment.[21]

à Terre-Neuve," both in Martijn (ed.), *Les Micmacs et la Mer* (Montréal, 1986), 163-194 and 197-224; and Olive Patricia Dickason, *Louisbourg and the Indians: A Study in Imperial Race Relations, 1713-1760* (Ottawa, 1976), 46. For scepticism that Newfoundland was part of the Mi'kmaq domain, see Ingeborg Marshall, "Beothuk and Micmac: Re-examining Relationships," *Acadiensis*, XVII, No. 2 (1988), 52-82; and Leslie F.S. Upton, *Micmacs and Colonists; Indian-White Relations in the Maritimes, 1713-1867* (Vancouver, 1979), 64. Significantly, archaeological work in the vicinity of Burgeo suggests there was no Recent Indian presence (including Mi'kmaq) in southwestern Newfoundland until after a seasonal European occupation introduced the attraction of "iron and other European materials;" Tim Rast, M.A.P. Renouf and Trevor Bell, "Patterns in Precontext Site Location on the Southwest Coast of Newfoundland," *Northeast Anthropology*, No. 68 (2004), 41-55, esp. 51.

[19]Mirliguèche was on the west coast of Lake Bras d'Or, Cape Breton Island. The mission never developed into a farming settlement but functioned instead as a Mi'kmaq assembly point on various occasions throughout the year, and sometimes as a place to leave the elderly, the women and the children when the men went on hunting expeditions; see Dickason, *Louisbourg and the Indians*, 72. In 1750, Abbé Maillard moved the mission to Chapel Island (Potloteg), near Port Toulouse. See Martijn, "Eastern Micmac Domain," 220-221; and Upton, *Micmacs and Colonists*, 34. Thereafter, the Mi'kmaq congregated annually at Chapel Island to celebrate the day of their patron saint, Anne.

[20]Upton maintained that the gifts were initially a form of rent, an acknowledgment by the French that they were intruders in the region, and that later the French regarded the gifts as "a form of retainer for future services;" *Micmacs and Colonists*, 36-38.

[21]Both Martijn and Dickason stress the degree of Mi'kmaq dependency upon French gifts, but Dickason also emphasizes the reciprocal nature of this dependency.

During the eighteenth century, the Mi'kmaq, like native people elsewhere in North America, took as much advantage of this situation as they
could. Wherever they were in a position to influence the balance of power between the French and the English, natives were able to increase their demands
upon the French for presents.[22] Failure to comply led to threats that they would
turn to the English. This was an empty threat, since local British officials who
tried to play this diplomatic game never seemed able to convince their superiors in London that gift-giving was more than a sort of unreciprocated generosity.[23] Nevertheless the French were careful to ensure that gift-giving occurred
at regular intervals in recognition of the great importance that natives, including the Mi'kmaq, placed on the process of renewing such relationships.

British military authorities who were stationed in Cape Breton Island
following its capture in 1758 may not have fully approved of the principle of
gift-giving – the military commander and governor of Cape Breton between
1758 and 1761, General Edward Whitmore, predicted that "this will be a Constant Annual Expence" – but they recognized that the tradition was so deeply
entrenched that it would be in England's best interests to maintain it.[24] Their
superiors in London, however, concluded that the elimination of the French
from North America had also eliminated the power balance which the natives
had so cleverly exploited, so that giving them gifts on a regular or frequent

---

Compare Martijn, "Eastern Micmac Domain," 221, with Dickason, *Louisbourg and the
Indians*, 87-88.

[22]Upton, *Micmacs and Colonists*, 31 and 40. In 1768, General Thomas Gage,
the commander in chief of British forces in North America, observed that "[the Indians'] Jealousy of our increasing Strength, and their former Experience, when the
French possessed Canada, have taught the Savages the Policy of a Balance of Power.
They had two Nations to Court them and make them Presents and when disgusted with
one, they found the other glad to accept their offers of Friendship;" Gage to Secretary
of State Lord Hillsborough, 12 March 1768, in Clarence Edwin Carter (ed.), *The Correspondence of General Thomas Gage with the Secretaries of State, 1763-1775* (2 vols.,
New Haven, 1931; reprint, Hamden, CT, 1969), I, 166.

[23]In a report on the Indian trade in North America, Capt. Gordon, Chief Engineer for North America, observed that the French system of trading with the Indians
was far superior to all existing British systems, and recommended its adoption by the
British. The report was forwarded by General Gage to Secretary of State Lord Shelburne on 22 February, 1767; Carter (ed.), *Correspondence of General Thomas Gage*, I,
121. See also Upton, *Micmacs and Colonists*, 36-39.

[24]TNA/PRO War Office (WO) 34/17 Amherst Papers, 98, General Whitmore
to General Sir Jeffery Amherst, 14 November 1760; and Julian Gwyn, "Edward Whitmore," *DCB, III (1741-1770)*, 662-663.

basis was no longer necessary.[25] In short, the English assumed that their relationship with the natives in Nova Scotia could be based on treaties, which need only to be arranged once, rather than on the renewal of relations through regular gift-giving ceremonies.[26]

While some Mi'kmaq bands deferred to this approach, for others the expectation of gifts died hard. This appears to have been the case with the Mirliguèche Mi'kmaq, led by Jeannot Pequidalouet (hereafter referred to as Jeannot), who negotiated a treaty with local English military authorities some time in late 1759 or early 1760, after news of the fall of Québec reached him.[27] Yet the treaty did not address all of the needs of Jeannot's people, with the result that they attempted to renew contact with the nearest available French, namely those in St. Pierre.

One of the more pressing Mi'kmaq concerns was their inability to secure the services of a Roman Catholic priest to attend to their religious needs.[28] By the eighteenth century Catholicism had become "an integral part" of Mi'kmaq identity, according to Upton, so that the death in 1762 of their only priest, Abbé Maillard, made the appointment of a replacement a matter of great urgency.[29] A despatch to the Board of Trade would later stress "The want of means to exercise their Religion they complain much of, and is the

---

[25]According to Dickason, General Amherst "ruled that Indians could no longer be supplied with arms and ammunition as it was no longer necessary to purchase their friendship or neutrality since the French had lost their footing in Canada. It would now only be necessary to keep the Indians aware 'of our superiority, which more than anything else will keep them in Awe, and make them refrain from Hostilities;'" Dickason, *Louisbourg and the Indians*, 88-89.

[26]Upton, *Micmacs and Colonists*, 37.

[27]In his letter to General Amherst of 1 December 1759, General Whitmore identified Jeannot as the chief of the Cape Breton Mi'kmaqs (TNA/PRO, WO 34/17, 46-47). Governor Montague Wilmot of Nova Scotia also referred to Jeannot as "chief of the Indians of the Island of Breton;" Wilmot to Board of Trade, 17 September 1764 (TNA/PRO), CO 217/21, 234. So did Samuel Holland, the Surveyor of Cape Breton Island in 1768; Holland, "A Description of the Island of Cape Britain," 1 November 1768 (TNA/PRO), CO 5/70, 14-45.

[28]Upton emphasized that in the immediate aftermath of the Halifax treaties of 1761, "the Micmacs were more interested in priests and presents than in land;" Upton, *Micmacs and Colonists*, 68.

[29]*Ibid.*, 33, 65 and 68; and Martijn, "Eastern Micmac Domain," 222.

cause of a communication they keep up with the Islands of Saint-Pierre & Miquelon, where they have recourse for Priests."[30]

Yet the British were not particularly sympathetic to this need. In part, this reflected the prevailing anti-Catholicism of eighteenth-century British officialdom; in part, it reflected persistent suspicion that Catholic priests had kept Anglo-French friction in Acadia active between 1713 and 1760.[31] The situation had been exacerbated further during the war, when news of Ternay's capture of St. John's in 1762 had triggered rumours of a Mi'kmaq revolt in support of a French military resurgence.[32] Whatever the reason, the English in Nova Scotia were in no mood to be assured by Mi'kmaq oaths of allegiance and signatures on treaties into providing the natives with presents and priests.

Rebuffed or ignored in this way by British authorities in Nova Scotia, Chief Jeannot by 1763 had begun to regard Newfoundland and St. Pierre as the solution to his people's needs. The Mi'kmaq already had some experience at hunting, trapping and fishing in Newfoundland – particularly on that part of the coast west of Fortune Bay where relatively few European fishermen were found.[33] The restoration of St. Pierre and Miquelon to the French in 1763 would permit the Mi'kmaq to resume contact with their former benefactors, allies and spiritual advisors. Yet this was not something that their former allies, the French, were willing to encourage; the last thing they needed in the

---

[30]TNA/PRO, CO 217/21, ff. 263, Benjamin Green to Board of Trade, 24 August 1766.

[31]TNA/PRO, CO 217/21, 312, Gov. Francklin to Board of Trade, 3 September 1766, and 179-180 and 234, Gov. Wilmot to Board of Trade, 17 September and 9 October 1765. General Amherst blamed the difficulty of negotiating a settlement with the Mi'kmaq on French priests. He complained to the Board of Trade in 1761 that the Mi'kmaq in the "north parts of this province [Nova Scotia and Cape Breton] had not yet wholly made their submission to His Majesty" due to the machinations of Jean Manach, a priest who worked among the Mi'kmaqs with Abbé Maillard and was imprisoned by the British authorities and sent to England in 1761; TNA/PRO, CO 5/60, 240, Amherst to Board of Trade, March 1761.

[32]Janzen, "French Raid," 45; and Upton, *Micmacs and Colonists*, 61.

[33]See Martijn, "Eastern Micmac Domain," 222-223; and Ralph T. Pastore, "Micmac Colonization of Newfoundland" (Unpublished paper presented to the annual meeting of the Canadian Historical Association, University of New Brunswick, June 1977), 9. French records recorded sixty Mi'kmaq families wintering in Fortune Bay in 1707 and 1708, while others lived temporarily on St. Pierre; Marshall, "Beothuk and Micmac," 57. Pastore and Martijn both suggest that game depletion on Cape Breton Island also contributed to Newfoundland's appeal. Martijn adds that "whenever survival was affected by various circumstances, the Eastern Mi'kmaq responded by shifting their activities to different parts of this vast island domain;" Martijn, "Eastern Micmac Domain," 225.

1760s was another irritant in the delicate relationship they had with the British.[34] It was certainly not something that the British were willing to permit, for they viewed the Mi'kmaq as a threatening element in a region where imperial authorities were anxious to nurture a British fishery – a perception reinforced in part by a number of incidents that had occurred in earlier decades.[35] They also suspected that the Mi'kmaq would provide the French with an opportunity to test the limits of British patience in the region near St. Pierre. Imperial authorities were therefore inclined to discourage the Mi'kmaq from establishing themselves as a permanent fixture in the region. The task of carrying out this objective fell to the only instrument of authority at their disposal – the officers and crews of the British warships stationed in Newfoundland.

The Events of 1763-1765

The first hint that the Mi'kmaq were relocating to Newfoundland came during the summer of 1763. Capt. Samuel Thompson, HMS *Lark*, had been instructed to patrol the west coast of the island of Newfoundland. Thompson was primarily concerned with ensuring that French fishermen confined their activities to the Treaty Shore, which at that time extended no further south than Point Riche. But when *Lark* appeared at the tiny settlement on Codroy Island off the coast of the extreme southwestern corner of Newfoundland, Thompson discovered that the inhabitants had been greatly intimidated by the unexpected appearance of Chief Jeannot and a number of Mi'kmaq to hunt and trap in the region. Jonathan Broom, the principal merchant there, requested and received

---

[34]Governor Wilmot of Nova Scotia would accuse the French at St. Pierre of engineering "the defection of several Indian families, to the amount of one hundred and fifty persons" from Cape Breton Island to St. Pierre in 1765; TNA/PRO, CO 217/21, 216, Wilmot to the Board of Trade, 17 September 1764. This view is not supported by the obvious nervousness of the French at having the Mi'kmaq come to St. Pierre; Ribault, "La Population des îles Saint-Pierre-et-Miquelon," 35-38.

[35]In 1727, Cape Breton Mi'kmaq seized a New England vessel at Port aux Basques; Archives des colonies, Paris (AC), C^{11}B, vol. 9, 50v-51 and 64-70v (copy kept by Fortress Louisbourg National Historic Park, Louisbourg, Nova Scotia [AFL]), Governor de Brouillan to Minister of Marine Maurepas, 13 September and 20 November 1727; and Great Britain, *Calendar of State Papers Colonial* (*CSPC*), vol. 35, No. 789, Lt. Gov. Armstrong to the Committee of Trade and Plantations, 17 November, 1727. In 1747/1748, when England and France were still at war, a party of forty Mi'kmaq over-wintering in southwestern Newfoundland plundered the homes of settlers and captured a number of prisoners, some of whom were put to death; Marshall, "Beothuk and Micmac," 65.

a small quantity of arms with which to defend the community.[36] Yet Jeannot assured Thompson that the Mi'kmaq meant no trouble to anyone, wishing only to buy a shallop with which to proceed to St. Pierre for the services of a priest. Jeannot used the opportunity of his meeting with Thompson to have a treaty of peace renewed – presumably the same treaty Jeannot had signed in 1759-1760 with the British military authorities in Cape Breton Island. He also gave Thompson a detailed request for cloth, kettles, gunpowder, shot, muskets, hatchets, shirts, hats, nets, fishing line, a boat compass and other items, all of which he expected to receive as gifts. Thompson affirmed the treaty and promised to pass the request for presents on to his superiors, but he was alarmed by their declared plans to head for St. Pierre. He therefore "forbad them positively" from establishing any contact with the French there "under pain of being carried Prisoners to the Governour if they were taken in ye attempt." Thompson conceded, however, that the Mi'kmaq would probably make the attempt anyway.[37]

Thompson's scepticism that the Mi'kmaq would adhere to his instructions was more a recognition of their needs as Roman Catholics than a conclusion reached because they were natives. He was equally mistrustful of the inhabitants of Codroy who, he suspected, were trading with the French at St. Pierre.[38] It was this trade rather than Mi'kmaq plans to settle in Newfoundland

---

[36]TNA/PRO, ADM 52/1316, No. 2, master's log, HMS *Lark*, 1763; and National Maritime Museum, Greenwich (NMM), Graves Papers (GRV), vol. 105, "Answers to Heads of Inquiry, 1763," esp. "Remarks" of Capt. Thompson and "Answers" of Capt. Thompson, No. 46.

[37]TNA/PRO, ADM 1/2590, No. 4, Thompson to Stephens, 16 April 1764. Governor Graves referred to it as "the Treaty with them made by Genl Whitmore;" NMM, GRV/106, Graves to Secretary of the Admiralty, 20 October 1763. A copy of the treaty evidently accompanied this letter but has since disappeared. Annotations indicate that Stephens passed Jeannot's request for presents on to the Board of Trade, though the Board had already received the request directly from Thompson; TNA/PRO, CO 217/20, part 2, 322, Thompson to Board of Trade, 28 April 1764. Martijn, citing Brown, claims that the presents for Jeannot and his band were subsequently delivered in 1764 by HMS *Tweed*; Richard Brown, *A History of the Island of Cape Breton* (London, 1869), 356-357, cited in Martijn, "Les Micmacs aux îles de la Madelaine," 271. But this was not the case. The Board of Trade advised the Admiralty and Captain Thompson that Jeannot's band fell under the jurisdiction of the Governor of Nova Scotia, and that the Indians should "apply" for presents to Halifax; TNA/PRO, ADM 3/72, 6, minutes of a meeting of the Lords of the Admiralty, 2 May 1764. Captains of Royal Navy ships cruising in Newfoundland waters were authorized to give Jeannot this message, as was Jonathan Broom; see Provincial Archives of Newfoundland and Labrador (PANL), GN 2/1A/3, 235, Palliser at York Harbour in the Bay of Islands to Jonathan Broom, 29 July 1764.

[38]NMM, GRV/105, "Remarks" of Capt. Samuel Thompson, HMS *Lark*.

which at this point most concerned Thompson and his immediate superior, Captain Thomas Graves, the commodore of the Newfoundland station in 1763.[39] Hugh Palliser, however, who succeeded Graves in 1764, was more alarmed by the implications of Mi'kmaq settlement in Newfoundland. He believed that their presence there not only placed British fishermen at risk but that close proximity to the French raised fears that a collaboration might ensue that would destabilize the region. In a despatch to the Board of Trade, he expressed his "Apprehensions of ye danger of permitting any Indians getting footing in this Country, as thereby the Fishery's will be in the same Precarious State as when the French and their Indians possess'd Placentia and the South Coast."[40]

Palliser's concerns were heightened by the discovery, shortly after his arrival in Newfoundland, that Jeannot and his companions had remained in southwestern Newfoundland, trapping furs through the winter. Though they returned to Cape Breton Island in the spring, Palliser was convinced that a permanent liaison was developing between the Mi'kmaq and the French at St. Pierre, and that this relationship would only intensify if allowed to continue.[41] Yet any attempt to deal with the Mi'kmaq was frustrated by the British authorities in Nova Scotia and Cape Breton Island, who seemingly encouraged the Mi'kmaq into migrating to Newfoundland. Those Mi'kmaq who came to Newfoundland in 1764 carried passes provided by British military authorities on Cape Breton Island.[42] Despite Palliser's complaints, the practice persisted through 1765.[43]

It is possible that some officials in Nova Scotia viewed Mi'kmaq migration to Newfoundland as a solution to one of their more troublesome problems. The Mi'kmaq were becoming restive in the face of the British refusal to help them secure the services of a Catholic priest, and in response to the Brit-

---

[39]NMM, GRV/106, Draft of Graves to Secretary Stephens, 20 October 1763.

[40]TNA/PRO, CO 194/16, 173-173v, Palliser to Board of Trade, 30 October 1765.

[41]TNA/PRO, CO 194/17, 26v-27, Palliser, remarks on Article 32 of his instructions.

[42]Palliser refers to this in a letter to Michael Francklin, the Lieutenant Governor of Nova Scotia, dated 16 October 1766; see TNA/PRO, CO 194/16, 308-309. Palliser indicates that he immediately wrote to Major Wallace, the commanding officer at Louisbourg, insisting that the passes be revoked.

[43]TNA/PRO, CO 194/16, 173-173v, Palliser to Board of Trade, 30 October 1765; and PANL, GN2/1A/3, 343, Palliser to Lt. Col. Pringle (Louisbourg), 22 October 1765.

ish rejection of gift-giving traditions long practised by the French.[44] Some senior officials, however, were convinced that Mi'kmaq contact with the French posed a security threat to Nova Scotia and shared Palliser's wish to have such contact cease.[45] Yet what could they do? As Governor Wilmot explained to the military commander at Louisbourg:

> ...thro' a decent Submission to the Authority of Government, [Jeannot] Applied for my leave to go over...for the purpose of trading and hunting; had I refused my Consent...he might have taken that liberty with impunity, nor indeed can I find out the Law which prevents any of the King's subjects passing from any part of this Dominion to the other...[46]

Wilmot was therefore not indifferent to Palliser's dilemma. If anything, he felt just as frustrated as Palliser by the persistence of Mi'kmaq movement across the Cabot Strait. He blamed the problem on the parsimony of the imperial authorities in London, who continued to deny the Mi'kmaq the presents they demanded. Wilmot warned the Board of Trade that "terms cannot be kept with the Indians...without incurring an expence to gratify their wants, and to prevent any disgusts arising from a neglect of them."[47] London's intransigence over the issue of gifts was driving the Mi'kmaq to Newfoundland and into renewing their old relationship with the French.[48]

---

[44]Pastore, "Micmac Colonization," 13, and Ralph T. Pastore, *The Newfoundland Micmacs: A History of Their Traditional Life* (St. John's, 1978), 14. See also Virginia P. Miller, "The Decline of Nova Scotia Micmac Population, A.D. 1600-1850," *Culture*, II, No. 3 (1982), 107-120, esp. 110.

[45]In 1766, Lt. Gov. Francklin urged Palliser to have the warships stationed near St. Pierre and Miquelon do more to prevent all contact between the Mi'kmaq and the French, for fear that such contact would "in time prove of very ill consequence to this Young Province as our settlements are very stragling and defenceless;" TNA/PRO, CO 194/16, 307-307v, Francklin to Palliser, 11 September 1766.

[46]Wilmot to Pringle, 12 December 1765, cited in Pastore, "Micmac Colonization," 13.

[47]TNA/PRO, CO 217/21, 216, Wilmot to the Board of Trade, 17 September 1764. Wilmot was reacting to the Board's refusal to comply with the request for presents which Jeannot had made in 1763 to HMS *Lark*. When Jeannot redirected his request to the authorities in Cape Breton, he was turned down by the commandant at Louisbourg.

[48]As Wilmot explained, "the same Indian chief mentioned by your lordships [i.e., Jeannot] had ineffectually applied to the officer commanding the troops at Louis-

Ironically, if anyone was more concerned than the British authorities in Newfoundland and Nova Scotia by the appearance of the Mi'kmaq in Newfoundland waters, it was the French. Even before the islands of St. Pierre and Miquelon were restored to the French in July 1763, the duc de Choiseul warned Governor d'Angeac not to permit the Mi'kmaq to visit the islands.[49] For one thing, the French had more pressing concerns on their mind as they re-established themselves in the fishery. Among the issues which subsequently generated a lively dialogue between d'Angeac and his British counterpart, Hugh Palliser, were French rights to send their warships to patrol the fisheries, the precise location of the boundaries between French and English fisheries, French trade with British colonials in America and French contact with the Newfoundland coast to cut wood, hunt and trade.[50] Many of these issues were inescapable consequences of the condition in which the French found the islands of St. Pierre and Miquelon upon their restoration in 1763. As one British naval officer explained, the islands were left "barren and desart...destitute of all the Necessarys of Life, without Materials for building Houses, or Provisions to support them thro the Winter."[51] The French really had no choice but to steal over to Newfoundland's south coast, particularly the more remote stretches of western Placentia Bay, the Burin Peninsula and Fortune Bay, where they hoped to trade with British settlers while avoiding notice by the authorities at Placentia.[52] Such trade, though illegal, was essential to the

---

bourg for some small allowance of provisions, and other necessaries, and the declarations he then made, of being obliged, on the refusal he met with, to have recourse to the Island of St. Peter, and I have lately had the mortification to find that he was not only well received there, but that he has continued on that Island ever since with his whole tribe." TNA/PRO, CO 217/21, 216, Wilmot to the Board of Trade, 17 September 1764.

[49]Choiseul's instructions to d'Angeac, 23 February 1763, cited in Ribault, "La Population des îles Saint-Pierre-et-Miquelon," 35.

[50]See Thorpe, "Debating Talents."

[51]TNA/PRO, ADM 1/482, 305, Capt. Lord Colvill to Admiralty Secretary Stephens, 25 October 1763.

[52]TNA/PRO, CO 194/15, 108, Thomas Graves to Board of Trade, 20 October, 1763. South coast residents were attracted to this trade because the merchants of St. Pierre sold them supplies and gear at half the price charged by British merchants in Newfoundland; TNA/PRO, CO 194/27, 122v-123, Palliser to Lord Halifax, 11 September 1765. A lively trade also developed between the mainland colonies of British America and the French. Lt. Dundas of *St. Lawrence* reported both New England and Nova Scotia traders at St. Pierre to his commander, Lord Colvill; TNA/PRO, ADM 1/482, 410-412, Colvill to Stephens, 13 November 1764.

French if they were to re-establish themselves quickly on St. Pierre. Governor d'Angeac did not need the additional aggravation of Mi'kmaq visitors from Cape Breton to draw the attention of the British to the region – as one of the French military officers explained, Mi'kmaq contact with the French could cause trouble that "could extend to us, given the disputes we had had before with our neighbours."[53]

But Mi'kmaq behaviour was guided by their own needs, not those of the French. Despite being rebuffed in St. Pierre, the pressure of changing circumstances in Cape Breton Island compelled the Mi'kmaq to begin moving in significant numbers to Newfoundland's south coast. By 1765, Chief Jeannot, together with roughly 130 to 150 of his people, were settling within the inner recesses of Bay d'Espoir, a body of water with several arms that extended deep into the Newfoundland interior. Palliser found this very disturbing. James Cook's survey work that year had suggested that Placentia Bay, Fortune Bay and Bay d'Espoir "terminate near each other, and almost in the center of the Island, from whence it's not above 2 or 3 days march down to the sea coast on either side of the island." Should the Mi'kmaq succeed in establishing themselves there, they would be able to command much of the island's interior.[54] Palliser therefore arranged for two of the smaller warships in his command – HM Sloop *Spy* and HM Schooner *Hope* – to overwinter in Newfoundland, one at Placentia, the other at the mouth of Bay d'Espoir. Thus, rather than escaping scrutiny, the Mi'kmaq had managed only to attract it.

To be sure, the decision to station *Spy* and *Hope* on the south coast through the winter was an expression of Palliser's concerted effort to crack down on a number of illicit activities in the region.[55] In 1765, the ships of Palliser's squadron – comprising Palliser's flagship *Guernsey*, frigates *Pearl* and *Niger*, and *Egmont*, in addition to sloop *Spy* and schooner *Hope* – all converged on the west side of Placentia Bay, not far from St. Pierre. The *Grenville* brig also served that summer in the region, supporting the cartographic survey of the South Coast under the direction of James Cook. *Hope* and *Niger* soon left for Labrador, as eventually did *Spy* and *Guernsey*, but not before

---

[53]Archives nationales de France, Colonies (AN Colonies), C12/2, ff. 22-22v (Library and Archives Canada [LAC], reel F-568); and AN Colonies C12/2, Baron de l'Espérance to the Minister of Marine, 28 April 1766.

[54]TNA/PRO, CO 194/16, 173-173v, Palliser to Board of Trade, 30 October 1765; and PANL, GN2/1A/3, 343, Palliser to Lt. Col. Pringle (Louisbourg), 22 October 1765.

[55]See, for instance, Whiteley, "Governor Hugh Palliser;" and Whiteley, "James Cook, Hugh Palliser, and the Newfoundland Fisheries."

Palliser had held court in Great St. Lawrence Harbour, hearing a number of cases involving illicit trade between local residents and the French.[56]

Yet as winter approached, *Spy*, Capt. Thomas Allwright, and *Hope*, Lieut. Stanford, returned to the south coast and made ready to over-winter there. It was a remarkable measure, for warships had never before wintered on the south coast. *Spy* arrived at Placentia in mid-November and resumed active patrolling in the vicinity of St. Pierre the following April, well before warships usually took station in Newfoundland.[57] This was consistent with the attention Palliser had been giving the French since 1764. *Hope*, however, took up station three hundred kilometres to the west of Placentia, in the shelter of Great Jervis Harbour, located at the mouth of Bay d'Espoir.[58] The distance from there to Pass Island on the far side of Hermitage Bay was about twenty kilometres. *Hope* was therefore to monitor movement in and out of Hermitage Bay, to discourage French visits to the area for timber and trade with Newfoundland residents, and to inhibit communication between the Mi'kmaq and the French at St. Pierre. Occasionally this entailed a short cruise, though given the season this was not without some risk.[59] Moreover, the abundance of fjords, inlets and coves in Bay d'Espoir and Hermitage Bay made it more sensible and effective to send patrols out in shallops – the workhorses of the fishery, which could be

---

[56]TNA/PRO, CO 194/16, 172v, Palliser to Board of Trade, 30 October 1764. He was not as optimistic that he could put a stop to the trade between the French and the residents of Newfoundland's south coast: "I am Satisfy'd that no Examples, Orders or threats will have any effect on such People as the Winter Inhabitants of Newfoundland are." Ship movements based on information in captains' letters and ships' logs, including TNA/PRO, ADM 1/2300, ix; ADM 51/674; ADM 51/4210, iv (*Guernsey*), ADM 51/674, ii (*Pearl*); ADM 51/4220 and ADM 52/1288 (*Hope*); and ADM 51/925 and ADM 52/4396 (*Spy*).

[57]TNA/PRO, ADM 51/925, and ADM 52/4396, captain's and master's logs, HM Sloop *Spy*. Indeed, in February, long before *Spy* could set out, Allwright began to send shallops out on patrols.

[58]TNA/PRO, ADM 51/4220, and ADM 52/1288, captain's and master's logs, HM Schooner *Hope*, Lt. [William?] Stanford; and ADM 1/2300, ix, captain's letters. Stanford took over command of *Hope* in October from Lt. John Candler. Great Jervis Harbour was also frequently identified as Grand Jervis Harbour.

[59]Shortly after Christmas, *Hope* was nearly wrecked in a snow squall while on a brief cruise across Hermitage Bay. Details in this and subsequent paragraphs are taken from entries for the period November 1765-March 1766 in TNA/PRO, ADM 51/4220 and ADM 52/1288, captain's and master's logs, HM Schooner *Hope*.

rowed or sailed, and which could therefore easily penetrate even the deepest recesses of Newfoundland's irregular coast.[60]

The merits of both the location and these activities were demonstrated almost immediately. *Hope*'s encounters with the natives began literally the day the vessel moored in Great Jervis Harbour – a shallop was found there with several Mi'kmaq men, women and children. Similar encounters occurred in the days that followed as *Hope*'s crew began to construct winter quarters on shore. In mid-December, a shallop with seventeen Mi'kmaq men, women and children was sighted; another was spotted shortly thereafter with twenty-nine on board, while two shallops with fifty men, women, and children were sighted on 18 December.[61] In many instances, the natives came to the schooner, possibly drawn by curiosity, possibly by an expectation of trade. On at least one occasion, in mid-January, a child was brought to the ship to be baptized, though it is not clear who would have performed the ceremony, as there is no evidence that *Hope*'s complement included a chaplain. Most likely the Mi'kmaq would have preferred that the child be baptized by a proper priest in St. Pierre – a Mi'kmaq shallop had appeared at St. Pierre on Christmas Eve in search of religious services, provisions and other forms of assistance.[62] But the French did not welcome the natives, and Governor d'Angeac's official posture remained decidedly cool to their visits.

It also quickly became apparent that others were in the area besides the Mi'kmaq. Several log entries reported shallops owned or manned by Englishmen – on one occasion, two natives appeared in English boats crewed by Englishmen. North Arm (or North Bay) in particular appeared to be a bustling hive of activity, with almost daily reports of shallops coming and going, both native and Englishmen, presumably for trade. Thus, at one point *Hope*'s crew

---

[60]A shallop was a large open boat, occasionally provided with a canvas shelter or partial deck, propelled by both sails and oars. It was quite common for shallops to be purchased or hired by the navy to expand the vessels available to the station for purposes of visiting outports or probing deep into bays like Hermitage Bay and Bay d'Espoir. They were typically given names that identified the home station of the warship from which came their crew and officer or even the warship itself; usually the single officer would be a junior lieutenant or even a midshipman. In addition to extending the patrolling capability of a warship, the shallops also gave midshipmen and junior lieutenants the kind of command experience that was an essential part of their training.

[61]TNA/PRO, ADM 52/1288, master's log, HM Schooner *Hope*, Mr. James Blake.

[62]Charles-Gabriel-Sébastien, Baron de L'Espérance, Governor d'Angeac's nephew and eventual successor, noted the arrival at St. Pierre of a shallop of "*sauvages micmacs*" from the Newfoundland coast on 24 December 1765; AN Colonies, C12/2, ff. 22-22v, Baron de l'Espérance to the Minister of Marine, 28 April 1766. See also Ribault, "La Population des îles Saint-Pierre-et-Miquelon," 35.

were the recipients of some venison, while on another occasion *Hope* sent its shallop into North Bay to secure a supply of nails. It is not clear who these English were, or what drew them there – the opportunity to trade with the Mi'qmaq or the opportunity to trade with the French. Whatever reason applied, what *is* clear is that all this activity conflicted with Commodore Palliser's determination to discourage completely any interaction between Newfoundland and the French islands of St. Pierre and Miquelon. And the natives who had gravitated to Hermitage Bay and Bay d'Espoir were only making the situation more difficult.

Certainly by 1766 Commodore Palliser used the Mi'kmaq presence in the region to paint an exceedingly black picture of the future of the fishery in order to secure support for his efforts from his superiors in London:

> ...those Indians...dispers'd themselves about the Country to the great terror of all our People in those parts, so that before the arrival of the Kings Vessels, they had begun to retire, & had determin'd to Abandon the whole Fishery to the Westward of Placentia Fort, for the Indians had already begun to insult and Rob them on pretence of want of Provisions; but under the protection of the Kings Ships, our People return'd and remain. The Chiefs of the Indians were Summon'd, and had deliver'd to them my Orders to quit this Country...[63]

It was therefore with renewed vigour that the ships, vessels and shallops of the Royal Navy resumed their efforts in the spring and through the summer and fall of 1766 to monitor shipping in and out of St. Pierre, seizing any vessels suspected of trading with the adjacent coasts of Newfoundland. Though *Spy* remained moored at Placentia until the beginning of April, it had three shallops under its wing that were constantly in and out of harbour, cruising, inspecting and occasionally seizing vessels accused of violating trade restrictions. By the time *Spy* added its weight to these patrols, *Hope* had also begun to cruise out to St. Pierre and Miquelon. Finally, in May, the rest of the station ships began making their appearance. As far as the Mi'kmaq were concerned, all this activity appears to have had the desired effect. By the time vessels were again stationed at Placentia and Great Jervis Harbour as winter approached, the better "to drive away the Indians and to keep the French off," as Palliser put it in a despatch, Mi'kmaq were no longer sighted or reported in Bay d'Espoir.[64] Instead, Palliser learned that they had moved over to the west

[63]TNA/PRO, CO 194/16, 302v, Palliser to Board of Trade, 21 October 1766.

[64]TNA/PRO, CO 194/16, 308-309, Palliser to Lt. Gov. Francklin (Halifax), 16 October 1766. Spending the winter at Placentia this time was the sloop *Favourite* (16); *Hope* schooner again wintered at Great Jervis Harbour. See TNA/PRO, ADM

coast of Newfoundland.[65] While some Mi'kmaq occasionally still made their way to St. Pierre, by 1767 they were no longer a disruptive presence in the region.[66]

Conclusion

Do the efforts of Commodore Palliser and the ships and vessels of the Newfoundland station deserve credit for discouraging the Mi'kmaq from restoring their link with the French through St. Pierre? Had the navy been successful in interdicting the migration of Chief Jeannot's band of Mi'kmaq from Cape Breton Island to Newfoundland? Strictly speaking, the answer must be no. The Mi'kmaq did settle on Newfoundland, though they appear to have avoided Bay d'Espoir, at least for a while, after the false start between 1763 and 1766; instead, they moved to Bay St. George, far from St. Pierre and the commercial activity that centred on the French island, and therefore posed far less disruptive an element in Anglo-French relations. The decision to station HM Schooner *Hope* almost certainly served as a disincentive to a continuing Mi'kmaq presence in Bay d'Espoir. While *Hope*'s crew took no steps to drive the natives away, the constant naval presence in the area did disrupt trade to such an extent that any contact between St. Pierre and the Mi'kmaq in Bay d'Espoir was sporadic at best. Contributing to Mi'kmaq frustration would have been French reluctance to welcome them at a time when the need to re-establish a French presence in St. Pierre required a more circumspect relationship with the English. In the final analysis, the Mi'kmaq found themselves in a region where English and French imperial friction was sufficiently intense to interfere with their efforts to pursue their own priorities.

51/347, captain's log, *Favourite*; and TNA/PRO, ADM 51/4220 and ADM 52/1288, captain's and master's logs, *Hope*.

[65]TNA/PRO, CO 194/16, 305-305v, Palliser to Board of Trade, postscript of 27 October 1766. James Cook "found...a Tribe of Mickmak Indians" in Bay St. George while conducting his survey of that coast in 1767; TNA/PRO, ADM 52/1263, 233, master's log, *Grenville*, 20 May 1767.

[66]A shallop of Mi'kmaqs arrived in St. Pierre in 1769 on the pretext of seeking news about the health of the French King and to assure the local authorities that their loyalty to France remained strong; Ribault, "La population des îles Saint-Pierre-et-Miquelon," 35.

The Royal Navy and the Defence of Newfoundland during the American Revolution[1]

When the War of American Independence began in 1775, one of the many dilemmas facing the Americans was how to make the British conscious of their threat. It seemed inconceivable to most British political and military leaders that the suppression of a colonial revolt would require a very determined or prolonged military effort. In 1774, Secretary of War Lord Barrington even predicted that, in the event of an American rebellion, the army would not be needed. "A Conquest by land is unnecessary," he explained, "when the country can be reduced first to distress, and then to obedience by our Marine." Lord North echoed Barrington's perception in 1775, although he conceded "that a Large land force is necessary to render our Naval operations effectual."[2] Few understood that the Patriot leadership enjoyed widespread sympathy and support, or that the Americans would be less concerned with trying to secure a military victory over the British than a political one. This entailed exerting sufficient pressure on the British government to cause it to abandon its efforts to crush the rebellion and accept instead a negotiated settlement. Towards this end, the political leadership of the American cause made the destruction of the British fishery at Newfoundland one of their earliest objectives. In so doing, they reminded the British that the stronger power did not necessarily have the ability to dictate the course of a war.

The Newfoundland fishery made an excellent target. It was widely regarded throughout the North Atlantic community as one of Great Britain's most important national assets. The wealth it generated was later estimated to have had a value in 1769 of £600,000, while the fishery's function as a "nursery for seamen" made it, according to the conventional wisdom of the day, an essential component of British seapower.[3] To ensure that the commercial and

[1]This essay appeared originally in *Acadiensis*, XIV, No 1 (1984), 28-48.

[2]Barrington is cited in Paul Kennedy, *The Rise and Fall of British Naval Mastery* (London, 1976), 114; and Lord North to William Eden, 22 August 1775, in William Bell Clark, *et al.* (eds.), *Naval Documents of the American Revolution (NDAR)* (12 vols., Washington, DC, 1964-), I, 684.

[3]The estimate of net gains from the fishery for 1769 is cited in Gerald S. Graham, *Sea Power and British North America* (Cambridge, 1941), 98. Compare this figure with that of £519,598, which is the calculated value of the codfish trade for the year 1770 according to Shannon Ryan, "Abstract of CO 194 Statistics," cited in Glanville Davies, "England and Newfoundland: Policy and Trade, 1660-1783" (Unpublished

strategic value of the fishery remained with the mother country, it was official British policy to preserve its migratory character.[4] The defence of overseas British possessions had traditionally been based upon the maintenance and exercise of naval power in metropolitan waters.[5] Because this conformed perfectly with British fisheries policy, hardly anything was done to provide Newfoundland with local defences. Not until 1770, when government approval was given to begin construction of new fortifications at St. John's, was a serious effort made to provide the fishery with an effective refuge in case of an attack.[6] But these works were far from complete in 1775, so that when American privateers and cruisers began to make their presence felt in Newfoundland waters in 1776, the fishery looked for its defence to the warships stationed there each year for the purpose of supervising its activity. To defend the fishery, dispersed along hundreds of miles of difficult coastline, was a formidable task which was further complicated by the fact that the Newfoundland stationed ships lacked both the numbers and the strength to protect the fishery against the American foe.

A demonstration of the threat which faced the British fishery at Newfoundland had already been provided late in the previous year. In November 1775, American privateers had disrupted the Canso fishery and then had attacked the Island of St. John, sailing boldly into Charlottetown harbour and plundering the town. It seemed only a matter of time before similar attacks would be made on the Newfoundland fishery as a means of exerting economic pressure on the British government. Indeed, in 1776 the Continental Marine Committee planned a major expedition to attack and destroy the fishery at Newfoundland. Upon learning of this plan, the Massachusetts state government

---

PhD thesis, University of Southampton, 1980), 328. On the perceived importance of the fishery as a "nursery for seamen," see Gerald S. Graham, "Fisheries and Sea Power," Canadian Historical Association *Annual Report, 1941* (Toronto, 1942), 24-31; and Jean-François Brière, "Pêche et Politique à Terre-Neuve au XVIIIe siècle: la France véritable gagnante du traité d'Utrecht?" *Canadian Historical Review*, LXIV, No. 2 (1983), 168-169.

[4]The circumstances leading to the adoption of the policy are outlined in C. Grant Head, *Eighteenth Century Newfoundland: A Geographer's Perspective* (Toronto, 1976), 35-41.

[5]Gerald S. Graham, "The Naval Defence of British North America 1739-1763," Royal Historical Society *Transactions*, 4th ser., XXX (1948), 95-97.

[6]Olaf Uwe Janzen, "Newfoundland and British Maritime Strategy during the American Revolution" (Unpublished PhD thesis, Queen's University, 1983), 104-105.

directed its armed cruisers to accompany the Continental warships.[7] In the end, the expedition never materialized, having fallen victim to a problem which bedevilled American efforts before 1778 to carry the war to sea, namely the lack of an effective navy. Continental and state governments experienced constant difficulty in competing with the owners of privateers for recruits and naval stores, because the latter could afford to pay higher prices, better wages, and larger shares of prize money.[8] Rebel governments were also frustrated by their inability to control the activities of the privateers.[9] As business ventures, privateers preferred targets which promised a maximum return for the least amount of risk and effort. Such concerns did not necessarily coincide with strategic requirements, as the repeated failure to mount an effective attack on the Newfoundland fishery during the war would demonstrate. Nevertheless, individual privateers did begin to make their appearance on the banks of Newfoundland shortly thereafter. Only the lateness of the season spared the fishery from serious damage.[10]

The naval and military establishment in Newfoundland did what it could to prepare for the expected onslaught, but this was never very easy. The squadron of warships stationed in the fishery, never very large in the best of times, had been steadily reduced in strength since 1769.[11] In that year it con-

---

[7]*NDAR*, VI, 271-273, Continental Marine Committee (Philadelphia) to Commodore Esek Hopkins, 22 August 1776.

[8]James C. Bradford, "The Navies of the American Revolution," in James J. Hagan (ed.), *In Peace and War: Interpretations of American Naval History, 1775-1978* (Westport, CT, 1978), 6.

[9]The raid on Charlottetown had been unauthorized; as George Washington complained in a letter of 20 November 1775: "Our rascally privateers men go on at the old rate, mutinying if they cannot do as they please;" cited in John Dewar Faibisy, "Privateering and Piracy: The Effects of New England Raiding Upon Nova Scotia During the American Revolution, 1775-1783" (Unpublished PhD thesis, University of Massachusetts, 1972), 44n.

[10]Great Britain, National Archives (TNA/PRO), Colonial Office (CO) 194/33, John Montagu to Lord George Germain, 12 November 1776.

[11]In this article the term "squadron" is frequently used when referring to the warships stationed at Newfoundland. Strictly speaking, this is not correct. A "squadron" was a detachment of warships which included several ships-of-the-line and was capable of exercising or disputing command of neighbouring waters. "Stationed ships" were smaller warships, usually frigates and sloops of war which escorted trade to a particular destination and then remained there on a temporary or seasonal basis; they were too weak to fight decisive engagements. See Graham, "Naval Defence," 96-97. Nevertheless, the term "squadron" has also been used in the broader sense of any small detachment of warships serving under the command of a single officer on a particular

sisted of one fourth-rate ship, three frigates, two sloops of war and a few insignificant brigs and schooners. In the ensuing years, government commitment to fiscal restraint had whittled that number away so that by 1775 the squadron consisted of the *Romney* (50), *Surprize* (28), two sloops of war and some armed cutters and schooners.[12] Hardly sufficient to carry out peacetime responsibilities, this number of warships was completely inadequate for the additional responsibility of meeting the privateering challenge. Complicating matters was the weak state of the fixed defences on the island itself. The new works at St. John's were incomplete, while the fortifications at Placentia were "in a very ruinous State & unfit for defence."[13] The troops in garrison at these places were never very numerous – barely a hundred men in 1775 – and generally were the castoffs of their parent regiments. In commenting on the detachment of the Royal Highland Emigrants serving at St. John's from 1776 to 1778, one of the senior officers of the regiment conceded with deliberate irony that "They were certainly the worst we could find in the B[attalio]n when they were sent there. So that we can say with a great deal of truth they were picked men."[14] So long as the major centres of the fishery were incapable of defending themselves, the commanders in chief at Newfoundland were compelled to station their warships at the more important harbours instead of cruising the fishery in search of privateers. In 1776, only one warship of the six in the squadron was able to patrol at sea; the others were either detached temporarily to other duties, being refitted or assigned to protect Placentia or St. John's harbours.[15]

---

service. It is in this sense that the warships on station duty at Newfoundland can be referred to as a "squadron;" see for instance G.R. Barnes and J.H. Owen (eds.), *The Private Papers of John, Earl of Sandwich, First Lord of The Admiralty* [*Sandwich Papers*] (4 vols., London, 1932-1938), I, 179-181.

[12]TNA/PRO, Admiralty (ADM) 1/470, and CO 194/32; ADM 51/29, 581 and 950, captains' logs; and National Maritime Museum, Greenwich (NMM), Robert Duff Papers, vol. 5. The figures in parentheses immediately following the ships' names throughout the article refer to the number of guns with which the ship was armed. Thus, *Surprize* was a twenty-eight-gun frigate, and *Penguin* sloop (see below, footnote 30) carried ten cannons and ten swivels.

[13]Provincial Archives of Newfoundland and Labrador (PANL), Colonial Secretary's Records, GN 2/1, vol. 6, Robert Linzee to Montagu, 13 August 1776.

[14]*New York Historical Society Collections*, XV (1882), 449-450, Letter-Book of Captain Alexander McDonald of the Royal Highland Emigrants, 1775-1779, Alexander McDonald to David Hay, 7 September 1778.

[15]Barnes and Owen (eds.), *Sandwich Papers*, I, 192-193, Montagu to Lord Sandwich, 28 May 1776.

The weakness of the Newfoundland station during the early years of the war was both a reflection and a consequence of the general condition of the Royal Navy before 1778. A decade of determined efforts by successive British ministries to reduce the national debt had imposed drastic restraints on navy spending, not only on the maintenance of the fleet in reserve but also on the fleet in readiness. Between 1766 and 1769, the naval estimate had been slashed nearly in half.[16] This left the Royal Navy poorly prepared for the sort of conflict which unfolded after 1775. It was a war in which the navy would be expected to carry out several demanding responsibilities – maintenance of sufficient strength in European waters to guard against French and possibly Spanish intervention in the war, protection of British bases in America, support for British military operations in America and patrols of an extensive American coastline to interdict American supplies and shipping.[17] At the same time, the British government was reluctant to respond to the situation in America with overwhelming force, in the belief that a reconciliation with the rebellious colonies was still possible. No attacks on colonial shipping were permitted for the first six months of the war; letters of marque were not issued until April 1777; convoy procedures were not adopted until the summer of 1776; a general press was not allowed until 1778.[18] Consequently, the North American squadron, like that in Newfoundland, had to assume wartime responsibilities at peacetime strength. It was an impossible task, as successive commanders of the North American station regularly reminded the Admiralty.[19]

---

[16]Piers Mackesy, *The War for America, 1775-1783* (London, 1964), 170-171; and Gerald S. Graham, *The Royal Navy in the War of American Independence* (London, 1976), 5.

[17]David Syrett, "Defeat at Sea: The Impact of American Naval Operations upon the British, 1775-1778," in *Maritime Dimensions of the American Revolution* (Washington, DC, 1977), 14-15. On 18 September 1776, Vice-Admiral Lord Howe, the senior officer on the North American station, provided the Admiralty secretary with a thorough discussion of his squadron's many responsibilities in order to justify its inability to suppress the activity of the privateers: TNA/PRO, ADM 1/487.

[18]Syrett, "Defeat at Sea," 13-22; Mackesy, *War for America*, 170-172; and Graham, *Royal Navy*, 6-7.

[19]See, for instance. TNA/PRO, ADM 1/484, Thomas Graves to Philip Stephens, 22 June 1775; ADM 1/488, Lord Howe to Stephens, 10 December 1777; ADM 1/489, James Gambier to Stephens, 20 December 1778; and ADM 1/486, Marriot Arbuthnot to Stephens, 29 August 1779 and 30 September and 14 December 1780. See also Syrett, "Defeat at Sea," 17; Faibisy, "Privatering and Piracy," 57-58; and George Comtois, "The British Navy in the Delaware, 1775 to 1777," *American Neptune*, XL, No. 1 (1980), 14-16.

Insofar as the Newfoundland station was concerned, this meant that there was an intense competition with other squadrons and stations for the insufficient men and ships which became available. And, invariably, the Newfoundland station was forced to defer to the needs of more important operations. For instance, in 1776 the Admiralty's intention to send the Newfoundland squadron out to the fishery as early as possible was frustrated by manning delays at Portsmouth, where warships destined for New York and Québec were permitted to bring their complements up to strength first. And, while the paper strength of the Newfoundland station was greater by two frigates than it had been in 1775, its effective strength was no greater for most of the season. *Surprize* (28) and *Martin* (14) were temporarily detached to the expedition sent to lift the American siege of Québec, while the services of *Cygnet* (14) were lost, first to a desperately needed refitting and later when it fell in with a Halifax-bound troop convoy which ordered her to accompany them. The Admiralty did order Vice-Admiral Molyneux Shuldham of the North American squadron to direct *Fowey* (24), a frigate in his command, to join the Newfoundland squadron. This well-intentioned gesture was frustrated by *Fowey*'s participation in an expedition against the Chesapeake; the ship could not be made available until the season was practically over.[20]

The commander of the Newfoundland station from 1776 to 1778 was Vice Admiral John Montagu, the first officer of such senior rank to serve there.[21] The *London Chronicle* welcomed his appointment, describing him as an officer who was "experienced and active" – a reference to his three years' service as Commander in Chief of the North American station, where his enforcement of British trade regulations had been particularly vigorous, even heavy-handed.[22] Now, he applied himself to the task of improving the effec-

---

[20]TNA/PRO, CO 5/254, Germain to the Admiralty, 6 January 1776 (transcripts in Library and Archives Canada [LAC]); Barnes and Owen (eds.), *Sandwich Papers*, I, 191-192, Montagu to Sandwich, 10 March 1776; *NDAR*. IV, 919-920, Admiralty to Linzee, 17 February 1776 and Admiralty to Harvey, 17 February 1776; Barnes and Owen (eds.), *Sandwich Papers*, I, 192-193, Montagu to Stephens, 28 May 1776; ADM 1/471, Montagu to Stephens, 15 August 1776; *NDAR*, IV, 1047, Admiralty to Shuldham, 18 April 1776: *NDAR*, V, 344-346, Shuldham to Stephens, 2 June 1776; and *NDAR*, VI, 88-89, Hamond to George Montagu, 6 August 1776.

[21]Montagu served in North America from 1771 to 1774. An outline of his career is provided in *Dictionary of National Biography* (Oxford, 1917), XIII, 705-706. In addition to his command of the Newfoundland station, Montagu was appointed governor of the island. Until the British parliament finally recognized Newfoundland as a colony in 1824, it was customary for the officer commanding the stationed ships at Newfoundland to be appointed governor as well. See Frederick Rowe, *A History of Newfoundland and Labrador* (Toronto, 1980), chap. 9.

[22]*London Chronicle*, 5-7 March 1776, cited in *NDAR*, IV, 948.

tiveness of his command. With Admiralty permission, he began to replace his squadron's cutters, sloops, and schooners with larger vessels capable of facing the challenge of the privateers. Concerted efforts were made to accelerate construction of the works at St. John's and to effect temporary repairs sufficient to halt the further deterioration of those at Placentia. In this way, the ships assigned to protect those harbours might be released to patrol the fishery. But for all his efforts, Montagu was not rewarded with much success. Although damage to the fishery during his first year of command was much lighter than had "been naturally expected," this was hardly to the credit of Montagu's warships, which had made no contact whatsoever with enemy cruisers. Over the next two years, the privateers did become a serious problem for the bank fishery, and beginning in 1778, they extended their activities to the inshore fishery and outports of the south coast. Against this onslaught, the Newfoundland stationed ships seemed helpless. Only two privateers were taken in 1777, and only one in 1778, whereas one of Montagu's frigates was captured, and a sloop of war and an armed schooner were wrecked in separate incidents. Noting the success with which Montagu's warships captured three enemy merchantmen with valuable cargoes in 1777, the Admiralty suggested that such activity might be more appropriately directed towards the protection of the fishery.[23]

Montagu objected vigorously to such thinly veiled criticism. "I beg leave to observe to their Lordships," he wrote in a letter to Philip Stephens, the Secretary to the Admiralty, "that I never did give an order to any Captain under my Command but it was for the protection of the fishery."[24] Any failure on the part of his warships, he added, was an unavoidable consequence of their lack of proper support. Frequent appeals for reinforcement had little effect. Despite purchases of smaller warships, the Newfoundland station from 1776 to 1778 would show little increase in effective strength. Although Admiral Montagu had been provided with an increase in frigates in 1777, two were armed with only twenty guns each, while the capture of *Fox* (28) in June by the Continental frigates *Hancock* (32) and *Boston* (28) left the squadron no stronger than in the previous year. As for 1778, in some respects the Newfoundland station was even weaker than in 1777. For most of the season, it had fewer armed vessels, fewer sloops of war and only one additional effective frigate.[25]

---

[23]*NDAR*, IV, 990-994, Admiralty to Montagu, 23 March 1776; TNA/PRO, ADM 1/471, Montagu to Stephens, 5 May, 21 June, 27 July, 15 August, 2 September and 12 November 1776; and Janzen, "Newfoundland and British Maritime Strategy," 187-188.

[24]TNA/PRO, ADM 1/471, Montagu to Stephens, 5 May 1778.

[25]TNA/PRO, CO 194/33, Montagu to Germain, 12 November 1776; and ADM 1/471, Montagu to Stephens, 13 March and 15 June 1777, and 5 and 19 May and

In contrast, American privateers and warships during that same period cruised in what seemed to British officials in London to be embarrassing numbers. Vice Admiral Sir Hugh Palliser, one of the Lords of the Admiralty from 1775 to 1779, observed that "The escape of so many privateers of force from so great a fleet as we have in America to watch them, and the taking of the *Fox*, is very mortifying and disgraceful."[26] The commander in chief of the North American squadron at this time, Vice Admiral Lord Howe, pointed out in his defence that collecting accurate intelligence was easier said than done. Furthermore, he maintained that acting upon such intelligence with a squadron which was too small for the many tasks assigned to it, whatever Palliser thought to the contrary, was even more difficult.[27] Reinforcement of the stationed ships in North America and Newfoundland was essential, if only to keep up with the expanding activities of the Americans.

Adding to Montagu's woes was a shortage of naval stores and an absence of service facilities in Newfoundland which persisted throughout the war. Every new acquisition to the squadron was armed, manned, and suited with sails and rigging by borrowing from the larger warships. The lack of repair facilities was still another problem. Ships with foul bottoms could not be properly cleaned; damaged masts and spars were either given temporary repairs, or the injured vessel was sent to Halifax for proper repairs. Occasionally, one problem fortuitously solved the other. When *Proteus* (20) sloop of war arrived at St. John's in 1778 in an unserviceable condition, it was permanently moored in the harbour where it was slowly and steadily cannibalized by the other vessels stationed at St. John's. While *ad hoc* measures provided some measure of relief from the problems of inadequate supplies and service facilities, generally the collective strength of the Newfoundland squadron was reduced. Worn-out suits of sails, rigging and spars which might have been replaced, had instead to be conserved. Of necessity, aggressive pursuit of priva-

---

27 July 1778. W.H. Whiteley describes Montagu's squadron in 1778 as "more powerful than ever before;" William H. Whiteley, "Newfoundland, Quebec, and the Administration of the Coast of Labrador, 1774-1783," *Acadiensis*, VI, No. 1 (1976), 108. But the strength of the squadron was always changing and never reached its projected strength for that year: see Janzen, "Newfoundland and British Maritime Strategy," 228n.

[26]Barnes and Owen (eds.)*Sandwich Papers*, I, 233-235, Palliser to Sandwich, 22 July 1777.

[27]TNA/PRO, ADM 1/488, Howe to Stephens, 10 December 1777.

teers was avoided or discouraged. Such considerations provided sufficient frustration in themselves, without the addition of the disapproval of superiors.[28]

On the other hand, from the government's perspective, the hardships of the Newfoundland squadron were secondary to the navy's principal role in the war before 1778, which was to support the military effort in North America. The Newfoundland fishery might be regarded as an economic and strategic resource of the first order, but the way to protect it from American cruisers, it was felt, was to bring the war in North America to a victorious conclusion. It was to this end that a strategy of military re-conquest in America was directed in 1777 and that naval stores and ships released from service in British waters were usually sent to North America.[29] Therefore, until his circumstances improved, Montagu would have to protect the fishery as best he could, with what he had. When he returned to St. John's with *Romney* (50) early in May 1777, there were only two other warships on station. *Martin* (14) was a sloop of war which had spent the winter at St. John's and since early April had been cruising along the coast south of that harbour. *Spy* (14), also a sloop of war, had wintered at Placentia but now joined Montagu at St. John's. Four smaller armed vessels could not be fitted out immediately because of the absence of naval stores.[30] Upon the return to St. John's of *Martin* from its cruise, *Spy* was sent out to patrol the banks beyond Cape Race. As additional ships of the squadron arrived at St. John's, they were sent out either to patrol the banks south of Newfoundland or to take up station to protect valuable centres of the fishery, such as Placentia. Thus, *Fox* (28) was ordered to patrol the Grand Banks between forty-two and forty-five degrees north latitude, while *Surprize* (28) was sent to protect Placentia until one of the armed vessels could be made ready and relieve the frigate.[31]

---

[28]TNA/PRO, ADM 1/471, Montagu to Stephens, 15 June and 2 August 1777; Janzen, "Newfoundland and British Maritime Strategy," 228n; Metropolitan Toronto Library (MTL), Letter-Books of Rear-Admiral Richard Edwards, 1779-1782 (Naval), Edwards to Charles Chamberlayne, 29 July 1779, and Edwards to Cadogan, 5 August 1780; and TNA/PRO, ADM 1/2307, iii, Sir Richard Pearson to Stephens, 31 August 1782.

[29]Graham, *Royal Navy*, 9; Barnes and (eds.), *Sandwich Papers*, I, 202; and Isaac Schomberg, *Naval Chronology; Or An Historical Summary of Naval & Military Events, From the Time of the Romans to the Treaty of Peace 1802* (4 vols., London, 1802; reprint, Charleston, SC, 2012), IV, 325-331.

[30]*Penguin* (10..10), sloop; *Postillion* (10), brig; *Egmont* (8), brig; and *Bonavista* (8), sloop.

[31]This would not be until mid-June; TNA/PRO, ADM 1/471, Montagu to Stephens, 12 November 1776 and 15 June 1777; ADM 51/581, "Log of Captain Henry

The deployment of Montagu's warships was therefore essentially a defensive one to discourage the harassment of the bank fishery, to watch over passing trade and to protect the major centres of the fishery as effectively as possible. To attempt to do more, such as to seek out and suppress the activity of American cruisers, was not realistic, even if the Newfoundland station were significantly reinforced. A successful campaign against American privateers would first require that their home ports be captured or blockaded. But since the North American squadron at this time was concentrating instead on its role of supporting the operations of the British army in America, the American home ports were watched poorly, if at all. For Montagu to attempt to suppress enemy activity in Newfoundland waters would have been like attacking the upper branches of a tree while the roots were left alone.

Consequently, Montagu carefully avoided a deliberate attempt to hold the American cruisers in check. Instead, he instructed his ships to patrol as regularly as possible in the cruising grounds favoured by the Americans; by remaining visible, they could discourage enemy activity – at least, that was the plan. The essence of Montagu's complaints was not that he lacked the means to seek out and destroy the enemy cruisers, but rather that he lacked the resources needed to defend his station from them. This became all too evident in 1777 and 1778. Following the capture of *Fox* in June 1777, Montagu rearranged his dispositions, pairing his ships up so that they could support each other and avoid a repetition of that humiliating loss. As a result, even less of the fishery could be covered by patrols than before, giving the Americans an additional advantage which they did not ignore. In 1778, Montagu attempted to concentrate his meagre resources in the area of the bank fishery, where the Americans seemed to be most active. This meant leaving the Labrador coast unprotected – a calculated risk which, as it turned out, proved very costly. The American privateer *Minerva* (24) chose that year to cruise and destroy the fishery of Labrador. At the same time, American privateers began focussing their attention on the inshore fishery and outports along the south coast; by one estimate, nearly two dozen fishing vessels were cut out of various harbours and several communities were plundered during the summer of 1778.[32]

Yet the year 1778 represented something of a turning point for the Newfoundland stationed ships during the War of American Independence. Thereafter it became apparent that the so-called "privateering menace" had been misunderstood almost from the beginning of the war. As the purpose be-

---

Harvey, HMS *Martin*;" and ADM 51/950,"Log of Captain Robert Linzee, HMS *Surprize*."

[32]TNA/PRO, ADM 51/950, II, "Log of Captain Robert Linzee, HMS *Surprize*;" ADM 5/5, VI, "Log of Captain William Williams, HMS *Active*;" ADM 1/471, Montagu to Stephens, 9 September 1777 and 2 September 1778; and CO 194/34, Pringle to Germain, 31 January 1779.

hind the presence of American cruisers in Newfoundland waters became clear, it was possible to develop a strategy to combat them which did not demand daily support of the fishery by the warships stationed at St. John's. This would release them for more traditional wartime activities such as protecting trade and watching for enemy descents upon the fishery such as had occurred in 1762. The most noteworthy development in 1778 was the French decision to intervene directly in the war. When the news arrived in Newfoundland, Vice Admiral Montagu immediately put into execution his secret instructions, which had been in his possession for several months, to capture the French islands of St. Pierre and Miquelon, a dozen miles off the south coast of the island. Although St. Pierre had occasionally provided refuge and intelligence for American privateers, its real importance to the Americans was in the movement of supplies and war materiel from France to the United States. Capture of the French islands gave Montagu a badly needed victory with which to restore his squadron's spirits.[33]

A more important consequence of French involvement in the war was that the Royal Navy could now be established on a proper war footing. The surrender of General John Burgoyne's army at Saratoga in October 1777 had discredited the strategy of military re-conquest which the British had been pursuing in America and raised interest in adopting a policy of maritime pressure and blockade as an alternative. At the same time, Saratoga seemed to make a French declaration of war inevitable.[34] As a result, the navy experienced a rapid expansion after December 1777.[35] Yet the Newfoundland station did not benefit immediately from this growth. What Montagu needed most were frigates, which were in extremely short supply. In August 1777, there had not been a single frigate available to send to sea with the Home Fleet, which Lord Sandwich regarded as England's only defence against invasion.[36] This meant that overseas stations, particularly those of secondary importance such as New-

---

[33]TNA/PRO, CO 194/23, Germain to the Admiralty, 30 April 1778 (secret), and William Knox to Montagu, 1 May 1778; CO 194/34, Montagu to Germain, 5 October 1778; and Janzen, "Newfoundland and British Maritime Strategy," 244. St. Pierre is linked with the activity of American privateers in Newfoundland waters in Gordon O. Rothney, "The History of Newfoundland and Labrador 1754-1783" (Unpublished MA thesis, University of London, 1934), 245-246; and in Gerald S. Graham, *Empire of the North Atlantic: The Maritime Struggle for North America* (Toronto, 1950), 209.

[34]Barnes and Owen (eds.), *Sandwich Papers*, I, 327-335, Lord Sandwich to Lord North, 7 December 1777; and Mackesy, *War for America*, 154-156.

[35]Mackesy, *War for America*, 176.

[36]*Ibid.*, 174; and Barnes and Owen (eds.), *Sandwich Papers*, I, 235-238, Sandwich to North, 3 August 1777.

foundland, had to wait until sufficient frigates had been commissioned to meet the minimum needs of fleets at home and in the more important overseas theatres such as North America and the West Indies. Thus the 1779 squadron at Newfoundland was no larger than that of the previous year, whereas by 1780 and 1781, the number of frigates in the squadron had very nearly doubled, and by 1782, there were thirteen warships on the Newfoundland station, of which six were frigates carrying thirty-two or more guns each.[37]

Also in 1778, the new fortifications at St. John's were sufficiently close to completion that they could at last be provided with a garrison of more than 400 men. Although the ordnance for the works did not arrive until 1779, temporary batteries using cannon borrowed from the old fortifications were laid, placing the harbour in an unprecedented state of defence. Thereafter, the strength of the harbour defences was steadily increased through continued improvements to the works themselves, and through the establishment of batteries at neighbouring coves and harbours as well as on the roads leading into the town.[38]

But of all the precautions and measures taken to strengthen the posture of defence at St. John's, none was quite so extraordinary as the creation in 1780 of the Newfoundland Regiment. This innovation was the climax of several years of effort by Captain Robert Pringle of the Royal Corps of Engineers, who was in charge of the construction of the new defences for the harbour at St. John's. He had long dreamed of establishing a corps of light infantry on the island of Newfoundland as the fishery's principal defence. He first explained his ideas in a plan submitted to the Board of Ordnance in 1773. Pringle reasoned that the fishery at Newfoundland would best be defended by the establishment of harbour batteries at the major fishing centres and by the creation of a corps of light infantry or "rangers" who would go to their support in case of an attack. The rangers would be recruited in Newfoundland. This would give them a familiarity with the terrain and conditions which would enhance their effectiveness and therefore multiply their strength. If garrisoned somewhere in the interior and linked with the major fishing centres by a network of woodland trails, they would be able to respond quickly and in over-

---

[37]Compiled from various documents in TNA/PRO, ADM 1/471 and 472. In 1777, the Royal Navy had eighty-nine frigates (including fifty-gun ships) in service; in 1778 that number had increased only to ninety-six, but by 1783 there were 180 frigates in service; see Graham, *Royal Navy*, 9n; and Antony Preston, David Lyon and John H. Batchelor, *Navies of the American Revolution* (Englewood Cliffs, NJ, 1975), 146.

[38]TNA/PRO, CO 194/34, Montagu to Germain, 30 July 1778; and PANL, GN 2/1/7 (Letters), Lieut. Col. Hay, "Present State of the Forts and Batterys in the Harbour of St. John's..." A more thorough discussion of the measures taken to secure and defend the harbour at St. John's during the war is provided in Janzen, "Newfoundland and British Maritime Strategy," chap. VII.

whelming strength to any appearance by hostile forces on the coast.[39] It was a plan with calculated appeal, since it offered an inexpensive if unorthodox defence based upon sound principles of mobility and concentration. Moreover, Pringle's recommendations concerning harbour batteries coincided with suggestions submitted to the government about the same time by Governor and Commander in Chief of the Newfoundland station, Rear Admiral Molyneux Shuldham, with whom Pringle had consulted in 1772. The government was receptive to such ideas since orders were given in the spring of 1773 to improve the security of the harbours at St. John's and Placentia with additional batteries.[40]

Pringle's concept of a ranger force was never endorsed, for reasons which can only be surmised. According to government perception and official policy, the Newfoundland fishery was a migratory one, based in England, whose defence was provided through the exercise of British sea power in metropolitan waters. Local defence served no other purpose than to provide a protected refuge in which the fishery might seek shelter in case of an emergency and only for as long as it took to send relief from England. This resulted in the conclusion, articulated in 1766 by Secretary of State the Duke of Richmond, that "the protection of the Inhabitants settled on the Island is neither practicable nor desirable."[41] Towards that end, government in 1770 had given its approval to construct new fortifications at St. John's which were to be modest in scale and would require no more than 300 men in garrison.[42] Pringle's plan, which assumed that the defence of the fishery required the defence of the major settlements on the island of Newfoundland, conflicted with that policy, and so it was ignored by the government.

---

[39]NMM, Robert Duff Papers, vol. 8, Pringle to the Board of Ordnance, April 1773, "A Plan for the General Defence of...Newfoundland."

[40]LAC, Dartmouth Papers (DP), series 1, Shuldham to Lord Hillsborough, 30 September 1772, Dartmouth Papers, transcripts of original manuscripts, 1713-1798, Vol. 12, No. 2429; TNA/PRO, CO 199/17, Shuldham, "Remarks...made...to His Majesty's Instructions...1772 & 1773;" CO 5/161 (transcripts), Lord Dartmouth to Lord Townshend, 15 April and 20 February 1773 (LAC); and TNA/PRO, War Office (WO) 55/1557, II, Pringle to Board of Ordnance, 1 March 1774. See Janzen, "Newfoundland and British Maritime Strategy," chap. IV, for a more detailed discussion of government thinking and decisions concerning local defence at Newfoundland during the period from 1770 to 1775.

[41]William L. Clements Library, Ann Arbor (WLCL), Shelburne Papers (SP), vol. 86, Duke of Richmond to Captain Hugh Debbieg, Royal Engineers, 28 June 1766.

[42]TNA/PRO, CO 104/29, Board of Ordnance to Hillsborough, 25 May and 13 September 1770; and Janzen, "Newfoundland and British Maritime Strategy," 104.

Nevertheless, despite his inability to convince his superiors that his ideas had merit, Pringle seized every opportunity to promote defensive measures which applied his ideas, if only in part. He was responsible in 1777 for organizing his workmen into a defence force, and then expanding this idea in 1779 into a civilian volunteer defence force of 360 men, consisting largely of the servants of local merchants and fishermen. To proceed from there to the creation of a provincial regiment in 1780 was a relatively simple step, especially since the governor by then was Rear Admiral Richard Edwards, who had experimented with similar measures when he had served as governor of Newfoundland during the Seven Years' War.[43]

While the posture of defence at St. John's was improved in this manner, the British government had given its permission to arm the outports as well, so that they might be able to resist occasional attacks by privateers. Beginning in 1778, the government made several hundred stand of small arms available for distribution in response to requests from merchants who had invested in the northward expansion of the fishery as far as Labrador. At the same time, Governor Montagu encouraged the more influential residents of the leading outports to support the erection of small batteries for the defence of their harbours, using ordnance no longer needed at Placentia. Montagu was probably responding to the advice of Captain Pringle, who had proposed such an idea late in 1777 after the governor had already departed for England.[44] With only a few exceptions, the response to these measures was favourable, so that by 1782, fishing centres from the Burin Peninsula around the Avalon Peninsula and up to Labrador were able to put up some resistance to American privateers. The measures seemed to have the desired effect. In the remaining years of the war, only one of the outports which had been so equipped was actually attacked by a privateer, and that attack was easily beaten off.[45] Yet only the largest communities had received weapons. To equip every fishing port with the instruments for its own defence was unrealistic; there were sim-

---

[43]TNA/PRO, CO 104/34, Pringle to Germain, 6 June 1778 and 4 February 1779; PANL, GN 2/1/8, Edwards to Pringle, 28 October 1779 (Orders); TNA/PRO, General Sir Jeffery Amherst Papers, WO 34, Pringle to Lord Amherst, 3 November 1779; and Frederic F. Thompson, "Richard Edwards," *Dictionary of Canadian Biography*, IV (Toronto, 1979), 259-260.

[44]PANL, GN 2/1/7 (Letters), Jeremiah Coghlan to Montagu, 10 September 1778; TNA/PRO, CO 5/261 (transcripts), Knox to the Master General of the Ordnance, 20 March 1779; CO 194/34, Germain to Edwards, 2 April 1779; PANL, GN 2/1/7 (Letters), Montagu to the merchants of Bay Bulls, etc..., 27 July 1778; and TNA/PRO, CO 104/34, Pringle to the merchants of St. John's, 17 November 1777.

[45]In September 1780, the American privateer *General Sullivan* (20) attempted to cut a vessel out of Trepassey harbour; TNA/PRO, ADM 1/471, Edwards to Stephens, 28 September 1780.

ply too many of them to be defended this way. Consequently, attacks on the outports of Newfoundland, particularly those along the south coast, would continue until the end of the war.[46] Arming the larger outports could do nothing to entirely discourage attacks on more vulnerable targets.[47]

What made these various provisions for the local defence of the fishery so extraordinary was not their effect, real or perceived, on the activity of the American cruisers. Rather, it was the way in which they departed from the policy which had governed local defence before 1778. Gone was the careful adherence to the principle that "the protection of the Inhabitants...is neither practicable nor desirable." The proliferation of outport batteries, the distribution of small arms, and especially the creation of the Newfoundland Regiment in 1780, which more than doubled the original intended size of the garrison, all suggest that the purpose for which the defences at Newfoundland had been established had changed considerably. So long as the Royal Navy could exercise command of the sea in European waters, the migratory fishery at Newfoundland was reasonably secure and stood in need of nothing more than a temporary refuge. The rebellion in America, with its recourse to commerce-raiding at sea, had exposed a flaw in this logic, since the British government could not respond effectively in American waters without weakening her strategic reserve at home. But the "Paltry Privateers," as Governor Edwards contemptuously referred to them in 1779, never threatened the survival of the fishery, at least not through their activities in Newfoundland waters.[48] It was only with the entry first of France, then of Spain and Holland, into the war that the ability of the British to exercise command of the sea became uncertain.

---

[46]"American Privateers... infest this Coast in great numbers, and have committed great depridations on the South and South West this Island;" PANL, GB 2/1, Books of the Royal Engineers, Correspondence, 1774-1779, vol. 1, Lt. Caddy to Board of Ordnance, 5 July 1779. In 1780, Edwards informed the Admiralty that there were "a number of Privateers of force being upon the Banks and Coast;" TNA/PRO, ADM 1/471, Edwards to Stephens, 13 August 1780. A similar observation was made two years later by Edwards' successor, Vice-Admiral John Campbell; ADM 1/472, Campbell to Stephens, 23 September 1782.

[47]In 1779, privateers plundered harbours at Fortune Bay, St. Lawrence and Burin; PANL, GN 2/1/7 (Letters), William Saunders to Edwards, 21 July 1779. A similar attack on Mortier in the spring of 1780 was frustrated only because an army officer happened to be in the village recruiting for his regiment and was able to organize a defence; TNA/PRO, CO 194/35, Edwards to Germain, 1 August 1780. A lightly armed privateer sailed into Twillingate harbour early in 1779 and plundered the stores before proceeding to Battle Harbour on the Labrador coast where it caused more damage; W. Gordon Handcock, "John Slade," *Dictionary of Canadian Biography*, IV, 713.

[48]TNA/PRO, CO 194/34, Edwards to Germain, 9 December 1779.

It was the threat from Europe, not that from America, which made the defence of the island of Newfoundland itself so necessary by 1780.[49]

Table 1
The British Fishery at Newfoundland during the American Revolution

	Quintals of Fish Made by			Quintals of Fish
	British Fishing	Bye Boat	Inhabi-	Carried to Foreign
Year	Ships	Fishery	tants	Markets
1772	305,391	155,847	298,605	481,347
1773	262,925	150,957	366,446	489,665
1774	237,640	159,525	230,540	600,220
1775	268,250	159,525	230,540	600,220
1776				
1777				
1778	80,000	215,300	205,840	386,530
1779				
1780				
1781	26,600	75,750	220,100	286,403
1782	25,300	60,400	214,350	306,917
1783				
1784	131,650	93,050	212,616	497,884
1785	170,373	111,994	262,576	606,276

Note: The accuracy of fishery statistics in the eighteenth century, always open to criticism and doubt, is never more in question than during the American Revolution. Statistics were compiled by the officers of the Newfoundland squadron as part of their peacetime function. The war interfered with this activity and prevented it altogether in certain years. Nevertheless, these are the only statistics we have, and as relative indicators of trends they are reasonably accurate. The term "British fishing ship" included the traditional migratory ship fishery, which was an inshore activity, and the bank fishery. The bye boat fishery was a branch of the migratory fishery. It consisted of small boat owners who migrated seasonally to the fishery as passengers on the fishing ships, but who left their boats and other equipment in Newfoundland. It, too, was an inshore fishery. See Shannon Ryan, "Fishery to Colony: A Newfoundland Watershed, 1793-1815," *Acadiensis*, XII, 2 (1983), 34-52; and Head, *Eighteenth Century Newfoundland*, 63 and 72-74.

Sources: This table is compiled from data in TNA/PRO, CO 194/21, and ADM 1/471; William L. Clements Library, Shelburne Papers, vol. 86; and Shannon Ryan, "Abstract of CO 194 Statistics" (Unpublished paper, Centre for Newfoundland Studies, Memorial University of Newfoundland).

---

[49]Expressions of concern for "the defence of the island," or words to that effect, can be found in TNA/PRO, CO 194/35, Edwards to Germain, 16 September 1780; in PANL, GN 2/1/9 (Letters), Edwards to Pringle, 22 October 1780; and in TNA/PRO, CO 194/35, Germain to Edwards, 16 March 1781.

Contributing to this development were changes within the British fishery at Newfoundland which were accelerated by the war. Throughout the eighteenth century, the fishing merchants of the English West Country had been reducing their direct activity in the fishery, preferring to concentrate on the trade. In so doing, they elevated the importance and stimulated the growth of a resident fishery. That fishery remained in the shadow of the migratory fishery until the War of American Independence. Then, the transatlantic movement of men and materiel which was the definitive characteristic of the migratory fishery was increasingly interrupted. The *coup de grâce* came in 1778, when the migratory fishery was required to fulfil its role as a "nursery for seamen;" as it lost its vital reserve of skilled labour to the navy, the migratory fishery went into a rapid decline, and was supplanted by the resident fishery.[50] By the end of the war, the resident fishery was supplying about seventy-five percent of the fish for the trade with southern Europe (table one). Since the government still regarded the fishery as a "most important Branch of the Nation's Commerce & Source of her Power," it accepted the need to protect "those very important Possessions of the Crown upon which it depends," namely the outports.[51] The retreat from the position which had been so neatly articulated by the Duke of Richmond only fifteen years earlier was complete.

For the naval garrison at St. John's, the implications of all of these changes were quite profound. So long as land defences had been fairly modest, limited in recent years to St. John's and Placentia, and were intended only to provide the fishery with a defended refuge in the event of an unexpected attack, the ships stationed in Newfoundland waters had played an important role in the local defence of the fishery. Once local defence was entrusted to the garrisons, fortifications and batteries at St. John's, Placentia and the major outports, the stationed ships began cruising farther and farther at sea. Whereas in 1776 they had cruised in waters no farther south than forty-two to forty-five degrees north latitude, by 1780 and 1781 they were regularly instructed to cruise as far south as thirty-eight degrees north latitude. This placed the Newfoundland stationed ships deep in the North Atlantic trade lanes. There they were expected to watch for any signs of a French expedition against Newfoundland. They were also expected to protect British trade while hunting for

---

[50]Head, *Eighteenth Century Newfoundland*, chap. 8; and Keith Matthews, "A History of the West of England-Newfoundland Fisheries" (Unpublished DPhil thesis, Oxford University, 1968), 416-417 and 464-484.

[51]TNA/PRO, CO 194/35, Germain to Edwards, 16 March 1781.

enemy ships and vessels – a rare example of strategic requirements coinciding with an opportunity for prize money.[52]

Soon there were handsome dividends in the number of enemy cruisers and merchantmen captured by the warships stationed at Newfoundland. In contrast to 1778, when only one privateer was taken, 1779 saw six privateers captured by the squadron, now under the command of Rear Admiral Richard Edwards. In 1780 the number was even greater, so much so that initial reports of Edwards' success were received with disbelief in England. Then came 1781, the Newfoundland squadron's very own *annus mirabilis*, when fifteen enemy cruisers were captured. Against this success, from 1779 to 1781 the squadron lost only two armed schooners and two sloops of war to the enemy. Admiral Edwards could be forgiven if, in reporting these results to his superiors, he disregarded the absence of any supporting evidence and confidently asserted that "our taking so many of the American Privateers and their disappointment in not Capturing the Quebec Vessels as they did last year, has distressed the Northern Rebels much."[53]

From these results it would be easy to conclude that the problem of protecting the fishery and trade at Newfoundland from enemy cruisers had been solved, and in spectacular fashion, by Rear Admiral Edwards.[54] In fact, such was not the case. Careful attention to the location and date of capture of these privateers confirms that few of them had been threatening the Newfoundland fishery when taken. *Rambler* (14) was captured in 1779 while *Sibyl* (28) was escorting trade to Portugal; *Venus* (16), *Independence* (16) and *Diana* (10) were all taken in 1781 as *Surprize* (28) and *Danae* (32) were returning from Halifax; *Montgomery* (13) was also taken in 1781 while *Maidstone* (28) was cruising off Cape Breton Island.[55] In short, most were taken in waters far be-

---

[52]MTL, Edwards Letter-Books (Naval), Admiral Edwards' instructions in 1780 to the commanding officers of some of his frigates: for example, Edwards to Captain Isaac Prescott, 28 July 1780.

[53]TNA/PRO, CO 194/35, Edwards to Germain, 28 September 1781. See also Janzen, "Newfoundland and British Maritime Strategy," 259-260; Historical Manuscripts Commission, *Fifteenth Report, Appendix. Part VI: Carlisle Manuscripts* (London, 1897), 442, Eden to Lord Carlisle, August 1780; and TNA/PRO, ADM 1/471, Edwards to Stephens, 20 August 1781.

[54]See for instance Rothney, "Newfoundland and Labrador," 250-255; Matthews, "History," 480; and Davies, "England and Newfoundland," 199-202.

[55]*Rambler* was taken 22 October 1779, more than 200 leagues from Cape Spear: NMM, Journals and Diaries, vol. 85, Captain Thomas Pasley, "Journal 1779-1780, HMS *Sibyl*;" TNA/PRO, ADM 51/442, VII, "Log of Captain James Wainwright, HM Sloop *Hinchingbrook*;" ADM 1/1014, No. 7, William Dickson to the Sur-

yond the coast of Newfoundland. Edwards' success was due more to the expansion of his warships' cruising range than to any increase of privateering activity within the fishery.

Indeed, under Edwards, patrols of the fishery itself had fallen off considerably. Warships were rarely sent north of St. John's except in response to specific complaints of occasional privateers. It was more usual for them to be sent south, escorting trade back and forth between St. John's and Placentia, then ranging out in search of privateers cruising the trade lanes adjacent to Newfoundland. Not infrequently, Newfoundland stationed ships were detached to escort trade bound for Halifax or Québec. At one point at the height of the 1781 fishing season, six of Edwards' warships were employed in this manner; a seventh was assigned to the Labrador coast. Of the four ships of force which this left at Edwards' disposal, three spent much of August cruising the outer reaches of the banks which, by that stage of the war, had been abandoned by the fishery.[56]

This does not mean that the fishery was neglected. By 1779 the commander in chief of the Newfoundland station had recognized that American privateers, in contrast to the American Patriot leadership, did not wish to make the Newfoundland fishery a primary target. Although they could be found cruising in Newfoundland waters, they did so in order to intercept the trans-Atlantic British trade, much of which passed by the island and which could provide the privateers with profitable prizes. The bankers which had been victimized at the beginning of the war and the shallops which were characteristic of the resident fishery were little more than targets of convenience. Plundering them of their gear, sails, rigging, stores, and even men, enabled the privateers to extend their cruising time in the trade lanes, thereby enhancing their chances of taking a truly valuable cargo. Significantly, it was the fishery on the south coast of Newfoundland, nearest the trade lanes, that was molested most frequently by the privateers; the fishery from St. John's to Fogo was rarely disturbed. Only once, in 1777, had the Americans managed to organize an expedition with the purpose of destroying the British fishery at Newfoundland. Consisting of the Continental frigates *Hancock* (32) and *Boston* (28), as well as a number of privateers, this expedition had captured *Fox* (28). Despite that success, the expedition could be regarded as a failure. The privateers disappeared almost immediately upon setting out, preferring to hunt more profitable prey, while the victorious frigates decided not to press their luck and turned instead for home. Although it raised a great panic within the Newfoundland fishery, the expedition had caused very little damage. With the exception of

---

veyor General, 10 September 1781; and ADM 51/572, IX, "Log of Captain William Parker, HMS *Maidstone*."

[56]MTL, Edwards Letter-Books (Naval), Edwards to Lloyd, 31 August 1780; to Prescott, 28 July 1780; to Lloyd, 23 July 1781; and to Keppel, 6 August 1781.

the south coast, the fishery at Newfoundland passed through the remaining years of the war relatively unscathed.[57]

The greatest damage to the Newfoundland fishery and trade did not occur at Newfoundland at all, but instead was inflicted in European waters. That was where the fishing fleets, assembling in the spring or converging upon southern European markets in the fall, were most attractive as targets and, coincidentally, were most easily detected and most vulnerable to attack. Enemy privateers swarmed out of French channel ports, hovered off the Irish coast, or cruised between the Azores and the Portuguese coast, knowing that all British trade (including the Newfoundland trade) must eventually pass by.[58] Moreover, that was where chance encounters with enemy fleets might occur with devastating effect. Such was the case in June 1782, when Cordoba's fleet of thirty-two ships of the line was on its way to a rendezvous with a French squadron and stumbled upon the Newfoundland convoy. Nineteen merchantmen were taken.[59] Even though privateers were known to be active in Newfoundland waters, trade approaching the island evidently felt that the worst danger had been left behind them. Thus, in 1779 Captain Thomas Pasley, HMS *Sibyl* (28), observed that the merchantmen under his escort from England to St. John's had "behaved uncommonly well till they were about three or four hundred Leag. to the westward of Scilly, when thinking themselves out of all danger...they thought my protection no longer absolutely necessary" and dispersed, making for their respective destinations within the fishery.[60]

As the so-called "privateering menace" became better understood, the possibility of a French raid upon Newfoundland loomed once again as the fishery's greatest perceived danger. Ever since the raid of 1762, Newfoundland's defences had been planned with the expectation that a similar attempt to de-

---

[57]See, for instance, PANL, GB 2/1/1, Caddy to Board of Ordnance, 5 July 1779; Gardner W. Allen, *A Naval History of the American Revolution* (2 vols., Boston, 1913; reprint, Williamstown, MA, 1970), I, 202-216; TNA/PRO, ADM 1/471, memorandum of Captain McBride, n.d. [Summer 1777?]; Janzen, "Newfoundland and British Maritime Strategy," 206-212; PANL, GB 2/1/1, Pringle to Board of Ordnance, 20 June 1777; and TNA/PRO, CO 194/33, Montagu to Germain, 11 June 1777.

[58]Janzen, "Newfoundland and British Maritime Strategy," 256-257; Patrick Crowhurst, *Defence of British Trade. 1689-1815* (Folkestone, 1977), 134; Dorset County Record Office, Dorchester [DCRO], Benjamin Lester's Particular Letter-Book, Benjamin Lester to Stephens, 20 November 1782; Graham, *Royal Navy*, 7-8; and Gomer Williams, *History of the Liverpool Privateers and Letters of Marque with an Account of the Liverpool Slave Trade* (London, 1897; reprint, Cambridge, 2011), 198.

[59]TNA/PRO, ADM 1/472, Campbell to Stephens, 24 June 1782; and DCRO, Lester's Particular Letter-Book, Lester to Preston, 19 October 1782.

[60]NMM, Journals and Diaries, vol. 85, Pasley, "Journal," 21 April 1779.

stroy the fishery would be made when England and France next found themselves at war with each other. That perception had occasioned, first, a reassessment of the fishery's defences during the 1760s, and then the decision in 1770 to begin construction of the new harbour defences at St. John's. By the time those defences neared completion in 1779, France and Great Britain were at war, so that practically every year between 1779 and 1782 saw alarms at St. John's in the belief that a French descent upon the fishery was imminent. Partly for this reason, improvements were immediately begun to give added strength to the new harbour defences. This required nearly three more years of work, but it made the harbour of St. John's much less dependent for its security upon the naval garrison stationed there.[61]

Several factors were therefore responsible for the decision to order the warships stationed at Newfoundland to patrol the outer edges of the fishing banks. The importance of the migratory fishery had declined. The principal outports were increasingly able to defend themselves against occasional visits by privateers. It was recognized that privateers were more interested in the passing trade than in destroying the British fishery at Newfoundland. The strength of the harbour defences at St. John's was unprecedented. And it was felt desirable to detect any approaching French squadron as early as possible. That the long-awaited attack never materialized was perhaps fortuitous, since the French did consider such an attack more than once.[62] In the final analysis, however, the French were more concerned with applying pressure on the British in the Caribbean and in Europe. Besides, their American allies wanted a share of the Newfoundland fishery when the war ended, and were suspicious of French designs on the island. Since the alliance with the United States was important to French war aims, the French government was unwilling to make the capture and occupation of Newfoundland one of its primary objectives.[63]

---

[61]Janzen, "Newfoundland and British Maritime Strategy," 260-281.

[62]Storm damage in 1779 prevented D'Estaing from conducting operations against British strongpoints in the North Atlantic, including Halifax and Newfoundland, as planned: Jonathan Dull, *The French Navy and American Independence; A Study of Arms and Diplomacy, 1774-1787* (Princeton, 1975), 160-161; and Barnes and Owen (eds.), *Sandwich Papers*, III, 122. In 1781, Barras wanted to take the French squadron then at Newport, Rhode Island, north against the Newfoundland fishery. The remonstrations of Generals Rochambeau and Washington persuaded him to participate instead in the campaign against General Cornwallis in Virginia: Mackesy, *War for America*, 349-350 and 413-414; Dull, *French Navy*, 222 and 239-242; and Howard C. Rice, Jr. and Anne S.K. Brown (eds.), *The American Campaigns of Rochambeau's Army 1780, 1781, 1782, 1783* (2 vols., Princeton, 1972), I, 39n.

[63]Janzen, "Newfoundland and British Maritime Strategy," 292-294; Dallas Irvine, "The Newfoundland Fishery: A French Objective in the War of American Independence," *Canadian Historical Review*, XIII, No. 3 (1932), 281-283; and Orville

The precautions taken on shore to protect the fishery against the attacks of privateers and the preparations at St. John's in anticipation of a French raid upon the island therefore made it possible for the commanders in chief of the Newfoundland station to send their warships to cruise the outermost edges of the banks. There they provided valuable service in the protection of transatlantic British commerce. This, in turn, enabled the Home Fleet and the North American stationed ships to husband their own meagre resources and to give more attention to other duties. Consequently, it is not enough, when evaluating the role of Newfoundland during the American Revolution, to state merely that Newfoundland remained aloof from strategic considerations.[64] The naval garrison at St. John's made a significant contribution to the security of British trade in its neighbourhood and indirectly to the execution of British maritime strategy during the War of American Independence. Moreover, the changes in official attitudes and actual measures for the defence of Newfoundland contributed to the gradual shift in British policy which marked the island's evolution from fishery to colony.

Murphy, "The Comte de Vergennes, the Newfoundland Fisheries, and the Peace Negotiation of 1783: A Reconsideration," *Canadian Historical Review*, XLVI, No. 1 (1965), 32-46.

[64]Gerald S. Graham, "Newfoundland in British Strategy from Cabot to Napoleon," in R.A. Mackay (ed.), *Newfoundland: Economic. Diplomatic. and Strategic Studies* (Toronto, 1946), 247.

The American Threat to the
Newfoundland Fisheries, 1776-1777[1]

During the War of American Independence, American privateers and warships caused grave concern among the fishermen, merchants and naval and military authorities at Newfoundland. Strenuous steps were taken to defend the fishery against their activities. The larger outports were provided with gun batteries, while British warships patrolled the fishery, capturing a few dozen American vessels, although not without suffering some losses of their own. Despite these efforts, many fishing vessels suffered at the hands of the privateers, especially in the opening years of the war. Sails were plundered; rigging, supplies and even men were taken; and the fishing vessels themselves were often destroyed. By 1778, the bank fishery virtually ceased to exist. It has therefore been reasonable for most historians to assume that Newfoundland and its fishery were favourite targets of the Americans, and that only through the determined efforts of British military and naval authorities, and with the cooperation of the people of Newfoundland, were they beaten back.[2]

This interpretation is now coming under scrutiny and revision. Instead of being the principal objects of privateer attention, it appears that fishing vessels and settlements were "targets of opportunity" as the Americans cruised the busy trade lanes linking the West Indies and North America with Europe. Supporting this contention is the fact that most American activity was reported to the south of St. John's, in those Newfoundland waters which lay adjacent to the transatlantic trade lanes. Hardly any privateers were reported to the north. Fishing vessels were meagre targets for privateers, who were intent on only one thing – capturing merchantmen with rich cargoes which would bring a profit to their crew and their owners. Plundering a fishing vessel rewarded the privateer with only one precious thing – time. The stores, gear and men taken

---

[1]This essay appeared originally in *American Neptune*, XLVIII, No. 3 (1988), 154-164. Dr. Janzen completed his graduate training in 1983 at Queen's University, Kingston, Ontario, specializing in the fields of naval and colonial history. An earlier version of this article was presented in 1985 at a seminar sponsored by the Maritime History Group in St. John's.

[2]See, for instance, Gordon O. Rothney, "The History of Newfoundland and Labrador, 1754-1783" (Unpublished MA thesis, University of London, 1934), 250-255; Keith Matthews, "A History of the West of England-Newfoundland Fisheries" (Unpublished DPhil thesis, Oxford University, 1968), 480; and Glanville Davies, "England and Newfoundland: Policy and Trade, 1660-1783" (Unpublished PhD thesis, University of Southampton, 1980), 199-202.

from the victim would permit the privateer to extend its cruise and improve its chances of capturing a truly worthwhile prize. The American threat to Newfoundland was therefore almost an incidental one.[3]

Nevertheless, it is also true that on at least one occasion Patriot leaders planned the destruction of the British fishery at Newfoundland, while less than a year later two Continental frigates actually did attack the fishery. The value of the Newfoundland fishery to England as a source of wealth and as a "nursery for seamen" was a truism of the eighteenth century. So, too, was the principle upon which Great Britain had based its defence of the fishery, that Newfoundland "will allwayes belong to him that is superior at Sea."[4] Since the fishery was supposed to be a migratory one based in England, and since any threat to the fishery came traditionally from England's European enemies, the British government had adopted a strategy of defence that emphasized strength and preparedness in European waters. Local defence at Newfoundland was minimal. To the rebel leadership, an attack upon the fishery therefore seemed not only an obvious means of causing injury to Great Britain but also one which carried with it a fairly good chance for success. Despite their intentions, however, the Americans never managed to do more than threaten the fishery. It is the object of this paper to examine and account for this disparity between the intent and the execution of American plans to destroy the British fishery at Newfoundland.

The Anglo-American colonies had a well-established seafaring tradition by 1775, a tradition which had served them well both in peace and in war. According to one authority, shipbuilding had become the largest single industry in the colonies by 1775; one-third of all vessels operating in the British empire were colonial-built, employing thirty thousand seamen. In times of war, some of the colonies would provide provincial naval forces or "sea militia," to protect local trade or even support British naval and military operations in the vicinity.[5] It was therefore both logical and consistent with the colonial maritime experience for the revolutionary leadership to establish a national

---

[3]Olaf Uwe Janzen, "The Royal Navy and the Defence of Newfoundland during the American Revolution," *Acadiensis*, XIV, No. 1 (1984), 28-48.

[4]Great Britain, "Report of the Committee of the Privy Council," 15 April 1675, in W.L. Grant and James Munro (eds.), *Acts of the Privy Council of England: Colonial Series* (6 vols., London, 1908-1912), I, 622.

[5]William M. Fowler, Jr., *Rebels under Sail; The American Navy during the Revolution* (New York, 1976), chap. 1; James C. Bradford, "The Navies of the American Revolution," in Kenneth J. Hagan (ed.,) *In Peace and War: Interpretations of American Naval History, 1775-1978* (Westport, CT, 1978; 2nd ed., Westport, CT, 1984), 3-26; and W.A.B. Douglas, "The Sea Militia of Nova Scotia, 1749-1755: A Comment on Naval Policy," *Canadian Historical Review*, XLVII, No. 1 (1966), 22-37.

navy with which to carry their struggle against Great Britain out to sea. The colonials, however, were ill-prepared to mount a united effort in 1775 against the foremost naval power in the world. Their seafaring tradition had always been limited to individual colonies; inter-colonial cooperation did not occur. Consequently, American efforts to organize a national navy after 1775 were tentative and fragmented. Recently, it has even been suggested that the Continental Navy might have been a mistake.[6]

From the start, the Americans fought not with one navy but with several. In 1775, George Washington's army maintained its own warships to capture desperately needed gunpowder from the British. Eleven states organized their own individual navies. The Continental Congress founded the Continental Navy late in 1775, although for many months it existed mostly on paper.[7] But by far the greatest degree of American maritime energies and resources was poured into privateering, a sort of legalized piracy in which privately owned, well-armed, heavily crewed vessels carrying official permission would attack enemy commerce for profit.[8] Because the enthusiasm for privateering further fragmented the ability of the Americans to carry the war to sea, perceptions of privateering during this period were mixed. Defenders of the practice stressed

---

[6]Jonathan R. Dull, "Was the Continental Navy a Mistake?" *American Neptune*, XLIV, No. 3 (1984), 167-170. See also William S. Dudley and Michael A. Palmer, "No Mistake about It: A Response to Jonathan R. Dull," *American Neptune*, XLV, No. 4 (1985), 237-243.

[7]Gardner W. Allen, *A Naval History of the American Revolution* (2 vols., Boston, 1913; reprint, Whitefish, MT, 2014), II, chap. 19, esp. 662; and Frank C. Mevers, "Naval Policy of the Continental Congress," in *Maritime Dimensions of the American Revolution* (Washington, DC, 1977), 3-11.

[8]William J. Morgan, "American Privateering in America's War for Independence, 1775-1783" (Unpublished paper presented at the Symposium on Piracy and Privateering sponsored by the International Commission on Maritime History, San Francisco, 1975); Philip C.F. Smith, "The Privateering Impulse of the American Revolution," *Essex Institute Historical Collections*, CXIX, No. 1 (1983), 48-62; Peter L. Wickins, "The Economics of Privateering: Capital Dispersal in the American War of Independence," *Journal of European Economic History*, XIII, No. 2 (1984), 375-395, esp. 377-378; and Carl E. Swanson, "American Privateering and Imperial Warfare, 1739-1748," *William and Mary Quarterly*, 3rd ser., XLII, No. 3 (1985), 357-382. William Falconer, *An Universal Dictionary of the Marine* (London, 1780), defined a privateer as "a vessel of war, owned and equipped by particular merchants and furnished with a military commission by the Admiralty, or the officers who superintend the Marine Department of a country, to cruise against the enemy, and take, sink, or burn their shipping or otherwise annoy them, as opportunity offers;" cited in Walter Minchinton, "Piracy and Privateering in the Atlantic, 1713-76" (Unpublished paper presented at the Symposium on Piracy and Privateering, sponsored by the International Commission on Maritime History, San Francisco, 1975.

the disruption which privateers caused to enemy trade, the cost to the enemy in terms of lost shipping and cargoes and the consequent dispersal of enemy naval resources as steps were taken to protect merchantmen from their attacks. Critics of privateering maintained that they consumed stores and materiel needed by the navy and which the navy would have put to better use. Privateers were also accused of exacerbating the navy's manning problems by luring men with higher wages and the temptation of prize money. Generally, however, privateers had long been accepted as an essential supplement to a weak navy.[9] It has been argued that during the War of the Austrian Succession (1739-1748), when the Royal Navy had been too heavily committed elsewhere either to check enemy commerce or to protect colonial trade as thoroughly as was desired, "privateering operations played the leading role in America's war effort and made a major contribution to British sea power by disrupting Spanish and French commerce."[10] It was therefore both logical and consistent with their own tradition for American colonials to respond with a vigorous revival of privateering when the American Revolution began a generation later. No less determined an advocate of a national navy than John Adams conceded that privateers had an important contribution to make to the cause of revolution.[11]

Yet if Adams was willing to admit that privateers had a role to play in the revolution, he remained adamant that the principal maritime weapon of the Americans ought to be a national navy. He and a few others believed that only such a navy could serve the express wishes and strategies of the American people, as defined by the revolutionary leadership. This navy need not be capable of standing up to the Royal Navy. Adams recognized that the success of the revolution depended upon the ability of the Americans to protect their trade and commerce with other nations. This, in turn, would depend upon secure harbours, rivers and coastal trade lanes. "To talk of coping suddenly with G. B. at sea would be Quixotism indeed," he explained to a colleague, "but the only Question with me is, can We defend our Harbours and Rivers? If We can We can trade."[12] Later, Adams would think of other uses to which a national navy might be put, such as harassing the enemy into concentrating its forces. Always, however, the object was the same: to keep oceanic trade lanes and

---

[9]Minchinton, "Piracy and Privateering," 327; and Bradford, "Navies," 6.

[10]Swanson, "American Privateering," 359; see also Capt. Peter Warren to Governor Clinton (New York), 6 July 1744, in Julian Gwyn (ed.), *The Royal Navy and North America: The Warren Papers, 1736-1752* (London, 1973), 32.

[11]Smith, "Privateering Impulse," 55.

[12]John Adams to James Warren, 7 October 1775, in William B. Clark and William J. Morgan (eds.), *Naval Documents of the American Revolution* (11 vols., Washington, DC, 1964-2012) (*NDAR*), II, 342.

American seaports open to commerce, thereby keeping the revolutionary war effort alive.[13]

The decision by Congress to create a Continental Navy was not made without considerable objection. There were strong misgivings about the expense that a fleet of larger ships would entail. In October 1775, Samuel Chase made the oft-repeated remark that the thought of a national navy was "the maddest Idea in the World...we should mortgage the whole Continent."[14] Apart from its cost, there were strong sectional suspicions that a large navy would mostly serve and benefit New England. Opposition to such a navy was therefore strongest among the representatives of the middle and southern colonies; Chase was himself from Maryland.[15] Sectional misgivings were exacerbated when the navy failed to carry out its first major mission, early in 1776, to clear enemy raiders out of southern waters. The squadron assigned to this task carried out a raid on the Bahama Islands instead, hoping thereby to secure gunpowder and munitions still needed by George Washington's army.[16]

Nevertheless, congressional enthusiasm for the navy grew as it took shape late in 1775 and early 1776. From its inception, it consisted of smaller warships, such as armed brigs, sloops and frigates. This reflected not just congressional penury but also a confirmation of that initial assumption, that the proper role for an American navy – coastal defence, commerce protection, communications and intelligence – was defensive and best served by such vessels. The first warships were merchantmen which had been purchased and refitted. These were to be followed by a dozen frigates which had been ordered by Congress late in 1775; these began sliding down the ways in May 1776. And, as the navy began to take shape and grow, so did interest in an offensive naval strategy.

It was argued that several advantages to the American war effort would ensue from such a strategy. Early in 1777, Robert Morris, one of the

---

[13]John Adams to James Warren, 19 October 1775, in *NDAR*, II, 528; and Bradford, "Navies," 6-7.

[14]Journal of the Continental Congress, 7 October 1775, notes of John Adams, in *NDAR*, II, 341n.

[15]Bradford, "Navies," 7; Fowler, *Rebels*, 56-57; "Commentary" by W.J. Morgan to Mevers, "Naval Policy," 23-24.

[16]Fowler, *Rebels*, 95. When Hopkins set out to attack the Bahamas, he had eight ships and vessels under his command – the frigates *Alfred* (24) and *Columbus* (24); the brigs *Andrew Doria* (14) and *Cabot* (16); the sloops *Providence* and *Hornet*; and the schooners *Fly* and *Wasp* – with about seven hundred men and over two hundred marines: *Hornet* and *Fly* collided and were forced to return to port before the attack could be carried out. The numbers in parentheses refer to the number of guns with which each vessel was armed.

members of the Marine Committee, discussed in a letter to John Paul Jones the
idea of attacking and perhaps seizing the British Caribbean island of St. Christopher. Apart from the war materiel which might be captured, such an attack
would compel the British government "to provide for the Security and protection of every Island they have & by that means they must divide their Force &
leave our Coasts less carefully gaurded [*sic*]..." Morris then addressed the
larger question of an offensive strategy:

> disturbing their Settlements & spreading alarms, Shewing &
> keeping up a Spirit of Enterprize, that will oblige them to de-
> fend their extensive possessions at all points is of infinitely
> more Consequence to the United States of America than all
> the Plunder that can be taken, if the[y] divide their Force we
> shall have elbow room & that gained we can turn about &
> play our part to the best advantage which we cannot do now,
> being constantly cramped in one part or another, It has long
> been clear to me that our infant Fleet cannot protect our own
> Coasts & that the only effectual relief it can afford us is to at-
> tack the Enemies defenceless places & thereby oblige them to
> Station more of their Ships in their own Countries or to keep
> them employed in following ours and either way we are re-
> lieved so far as they do it...[17]

An attack on the Newfoundland fishery was perfectly consistent with
this logic. According to the conventional wisdom of the day, the British fishery
at Newfoundland was an economic and strategic asset of the first order. It was
a source of national wealth by virtue of the employment it gave to thousands of
fishermen, the trades and industries it supported in Great Britain and the trade
in fish with southern Europe and the Caribbean which it supplied. It was also
reputed to be a training ground or "nursery" for seamen, transforming landsmen into experienced mariners – a most important function for a nation whose
principal instrument of war was its navy. Yet the fishery was lightly defended
in the belief that British sea power in European waters secured it from any
reasonable danger.[18] Thus, at the very least, an attack upon the fishery by the
Continental Navy would alarm the British and compel them to disperse their
forces at relatively little risk to the Americans. If it were to succeed, such an
attack would severely damage a major source of England's economic strength

---

[17]Robert Morris to John Paul Jones, 5 February 1777, in *NDAR*, VII, 1109-
1111; see also Bradford, "Navies," 13.

[18]Gerald S. Graham, "Newfoundland in British Strategy from Cabot to Napoleon," in R.A. McKay (ed.), *Newfoundland: Economic, Diplomatic and Strategic Studies* (Toronto, 1946), 245-264.

and sea power. There were also some specific advantages to be gained by an attack upon the fishery at Newfoundland. Captured fish could be exchanged with the French for products needed by the rebel war effort. This was especially appealing to the New Englanders, whose own fishery had been interrupted by the war with disastrous consequences to their export trade.[19] The interruption of the supply of fish to the West Indies, where it was used to feed the slave population, might further complicate Great Britain's ability to give the rebellion in the colonies its full attention. Finally, there was the emotional satisfaction such an attack would give the Americans. As a correspondent of John Adams explained, "the Poole Men, who meant us much Harm, will be rewarded according to their Deeds..."[20] Consequently, on 22 August 1776, Esek Hopkins, the commander in chief of the fleet, was instructed to attack and destroy the British fishery at Newfoundland.[21]

This, however, was easier said than done. It had been under Hopkins' leadership that the raid upon the Bahamas had been carried out. As they returned to American waters, Hopkins' ships had fallen in with HMS *Glasgow*, a British frigate which mauled the American vessels before making its escape. For most of the remainder of the year, the Continental Navy remained incapacitated by the need for repairs, deficiencies in arms and munitions and especially sickness and shortages of men. For instance, over two hundred men, too sick to serve, had to be landed at New London immediately upon their return from the Bahamas. Then the replacements, which Hopkins had obtained temporarily from the army in order to work his ships round to Providence, had to be returned. Sickness spread through his ships. By May, his crews were down to half-strength. Hopkins could send one or two of his ships out on short cruises only by drawing men from all of his vessels.[22]

Adding to his woes, Hopkins was ordered to appear before the Congress in August to defend his conduct generally, and in particular to explain his decision to attack the Bahamas rather than carry out his instructions to clear southern waters of commerce raiders.[23] John Adams, who spoke in Hopkins'

---

[19]John Adams to James Warren, 6 April 1777, in *NDAR*, VIII, 282.

[20]John Lowell (Boston) to John Adams, 14 August 1776, in *NDAR*, VI, 181.

[21]Continental Marine Committee to Commodore Esek Hopkins, 22 August 1776, in *NDAR*, VI, 271-273.

[22]Esek Hopkins (*Alfred*, New London) to John Hancock, 9 April 1776 and (from Providence) 1 May 1776, in *NDAR*, IV, 735-736 and 1358-1360; and same to same, 22 May 1776, in *NDAR*, V, 199-200.

[23]Hancock to Esek Hopkins, 14 June 1776, and to George Washington, 14 June 1776, in *NDAR*, V, 528-531.

behalf, charged that "the Commodore was pursued and persecuted by that Anti New England Spirit, which haunted Congress..." Adams attributed that spirit to "the Prejudices of the Gentlemen from the southern and middle States."[24] Hopkins, however, must have been expecting something like this; two months earlier he had confided to his brother Stephen, who was a member of the Marine Committee, that "The several difficulties that attend the Navy are too many to mention & perhaps imprudent to Name, it is too much for my Capacity to Surmount." Commodore Hopkins then indicated that, if it were thought necessary, he would step down from his command.[25] Congress evidently did not feel that so drastic a step was required. Hopkins was censured, but he retained his command. It was shortly thereafter that the decision was made to attack the British fishery at Newfoundland.[26]

According to his instructions, Hopkins was to prepare *Alfred*, *Cabot*, *Columbus* and the recently acquired brig *Hampden* for a six months' cruise to Newfoundland "with orders to destroy the British fishery there:"

> they must make Prize of every British ship or Vessel they meet with – they must seize and destroy their Fishing Boats and Stages and make Prisoners of their fisher Men, or such of them, as will not freely enter into our Service, and as it is highly probable they may take more Prizes than they can conveniently spare Men to bring into Port, it may be proper in such case to destroy them. The season is now come when the Newfoundland Men begin to load their Fish Cargoes, consequently no time must be lost...

Once the vessels had "done...as Much Mischief as they can" in Newfoundland, they were to cruise in the Gulf of St. Lawrence to intercept the trade making for Québec, and then hunt down the Hudson Bay Company fur ships.[27] It was an unequivocal departure from the defensive strategy which the navy's last mission had been intended to serve. Perhaps the new emphasis on offense was a deliberate attempt to soften Hopkins' censure by giving him the sort of

---

[24]Journal of the Continental Congress, 12 August and 16 August 1776, in *NDAR*, VI, 157, 157n, and 209; extract for 12 August 1776 from John Adams' autobiography, in *NDAR*, VI, 157.

[25]Esek Hopkins to Stephen Hopkins, 8 June 1776, in *NDAR*, V, 425.

[26]Journal of the Continental Congress, 16 August 1776, in *NDAR*, VI, 209n; and Esek Hopkins to Hancock, 17 August 1776, in *NDAR*, VI, 220, 220n.

[27]Continental Marine Committee to Commodore Esek Hopkins, 22 August 1776, in *NDAR*, VI, 271-273.

mission he seemed to prefer. At least one of the state governments seemed to approve. The Massachusetts Council ordered its own warships to prepare to accompany the expedition and even tried to convince privateers to participate in the belief that a larger force "may more effectually answer the Purpose of Harassing Our Enemies and Destroying their Fishery."[28] Thus, the planned raid on the Newfoundland fishery had enthusiasm as well as logic on its side. Yet neither were sufficient to overcome the persistent difficulties which Hopkins faced in getting his ships ready for sea.

For one thing, *Hampden* had only recently been purchased into the navy and was still being refitted as a vessel of war. Moreover, its hull was in desperate need of a cleaning, as was apparently also the case with *Alfred* and *Columbus*. A further complication was that both *Cabot* and *Columbus* were out on a cruise, *Columbus* not returning before the end of September, whereupon it was immediately sent into the graving dock. Hopkins therefore could never be certain when he would have enough ships to carry out his instructions. Construction of two new warships at Providence was expected to be finished by the beginning of October. These were the frigates *Providence* and *Warren*. But their addition to his little squadron would only worsen his most pressing problem, namely that of providing his ships with men. Throughout the summer, Hopkins had laboured in vain to bring his complements up to strength. By October, when he should already have been at sea, he was still short four hundred men for the new frigates as well as for *Alfred* and *Columbus*. At one point, *Alfred*'s crew had been reduced to thirty men when it should have had two hundred more than that.[29] By the middle of the month, with the expedition still in port because of manning difficulties, Hopkins was intimating that it was now too late to intercept the Newfoundland trade.[30]

To salvage something out of the shambles of his mission, Hopkins decided to reduce its scale and redefine its object. He offered the command of *Alfred*, *Providence* and *Hampden* to John Paul Jones, whose mission would still be "to Destroy the Fishery of Newfoundland – but principally to relieve an Hundred of our fellow Citizens who are detained as Prisoners and Slaves in the Coal Pits of Cape Briton." Jones was agreeable to the change in plan, but he remained "Under the greatest Apprehension that the Expedition will fall to

---

[28]Journal of the Massachusetts Council, 21 August 1776, in *NDAR*, VI, 248-250.

[29]Esek Hopkins to the Marine Committee, 1 September and 10 September 1776, in *NDAR*, VI, 639, 770; Esek Hopkins to Capt. Hoysteed Hacker (*Hampden*, New London), 9 September 1776, in *NDAR*, VI, 757; Esek Hopkins to Governor Jonathan Trumbull, 15 October 1776, in *NDAR*, VI, 1271; and Jones to Morris, 17 October 1776, *NDAR*, VI, 1303.

[30]Esek Hopkins to Hacker, 14 October 1776, in *NDAR*, VI, 1253.

nothing," given *Alfred*'s continued shortage of seamen.[31] As it transpired, only two vessels would finally sail on 1 November (*Alfred* and *Providence*), and they were content to cruise no farther than the waters of Nova Scotia where they took several prizes.[32] Several ships and vessels were captured in Newfoundland waters late in 1776, but they were taken by privateers who had nothing to do with the planned raid on Newfoundland and never threatened to destroy the fishery as had been intended by the Marine Committee.[33]

Neither Hopkins nor Jones had any doubts about the reason for the failure of the 1776 operation. Both men blamed the manning problem, which they attributed to the popularity of privateering. The difficulties they encountered in manning their warships could, in fact, be traced back to the beginning of the war, when thousands of mariners and seamen enlisted in the state militias and the Continental Army. But responsibility for the persistence of the problem was placed squarely on the shoulders of the privateers.[34] In particular, they blamed prize money, which was much more lucrative among the privateersmen than it was within the navy. The entire value of a prize taken by a privateer went to its owners, officers and crew. In the navy, half the value went to the captors if the prize was a warship; in all other cases, the captors received only a third.[35] Hopkins argued that if the distribution of prize money were improved, then "it would be a great deal easier to Mann the Continental Vessels."[36] Jones concurred, explaining that:

> It is to the last degree distressing to Contemplate the State and Establishment of our Navy. – The common Class of mankind are Actuated by no nobler principle than that of Self Intrest – this and this Only determins all Adventurers in Privateers; the Owners as well as those whom they Employ.

---

[31]Jones to Morris, 17 October 1776, in *NDAR*, VI, 1303-1304.

[32]Fowler, *Rebels*, 96-97.

[33]Excerpts from the *Public Advertiser* (London), 25 October and 29 October 1776, in *NDAR*, VII, 710-711; and Great Britain, National Archives (TNA/PRO), Colonial Office (CO) 194/33, Gov. J. Montagu to Sec. of State Lord George Germain, 12 November 1776.

[34]Jones to Joseph Hewes, 19 May 1776, in *NDAR*, V, 151-153; and Esek Hopkins to Hancock, 22 May 1776, in *NDAR*, V, 199-200.

[35]Gardner W. Allen, "State Navies and Privateers in the Revolution," Massachusetts Historical Society *Proceedings*, XLVI (1912/1913), 189.

[36]Esek Hopkins to the Marine Committee, 10 September and 22 September 1776, in *NDAR*, VI, 770, 949.

> And While this is the Case Unless the Private Emolument of
> individuals in our Navy is made Superiour to that in Priva-
> teers it never can become respectable – it never will be for-
> midable. – And without a Respectable Navy – Alas America!

Jones recommended that all prize money be distributed among the officers and crews of the warship responsible for the capture. The proportion reserved for the public treasury was not so great that the government would miss its ab-sence, yet it would go far to stimulate enlistment into the navy.[37] To a point, Congress agreed. In October 1776, the prize share which went to captors of enemy warships was increased to 100 percent, while their share of other ves-sels was increased to one-half.[38] It remained to be seen whether this would have the desired effect of reducing the navy's manning difficulties.

The popularity of privateering was undeniable. It has been estimated that Americans outfitted two or three thousand privateers during the war, car-rying eighteen thousand guns and crewed by seventy thousand men – and these figures are probably conservative.[39] At the very time in August 1776 that Con-gress approved the attack on Newfoundland, a merchant in Boston was re-marking upon the enthusiasm with which New Englanders took up privateer-ing. He declared that "the Success of those that have before Engaged in that Business has been sufficient to make a whole Country privateering mad."[40] Everyone seemed to prosper, even the lawyers who thrived on the legal work needed to adjudicate and process the prizes. "They are making Money as fast almost as they Can receive it," remarked one cynic to a lawyer, conceding that "I suppose there never was a better chance for Gentleman of your Profession getting money than the Present, Privateering prevails so much & such a num-ber of prizes are taken that it makes a vast deal of Business." He added that the demand to outfit privateers was so great "that Cannon cannot be procured, if at all, but at a most extravagant price."[41] It was a situation which could not but complicate the task of the Continental Marine Committee as it tried to pre-

---

[37]Jones to Morris, 17 October 1776, in *NDAR*, VI, 1303.

[38]Allen, "State Navies," 189.

[39]*Ibid.*, 185; Smith, "Privateering Impulse," 50-51; and Morgan, "American Privateering," 9.

[40]Joseph Warren to Samuel Adams, 15 August 1776, in Massachusetts His-torical Society *Collections*, 7th ser., *Warren-Adams Letters* (2 vols., Boston, 1917-1925), I, 438.

[41]Thomas Cushing to Robert Treat Paine, 9 September 1776, in *NDAR*, VI, 755-756.

pare its own ships for war. Within the Marine Committee, however, opinion on the issue of privateering was divided. John Adams noted approvingly how "thousands of schemes for Privateering are afloat in American imaginations." While many of these proposals would probably amount to nothing, nevertheless it was important to provide encouragement knowing that "some profitable Projects will grow."[42] Although anxious that the new Continental Navy should survive its birth and infancy, he was convinced that privateering was also important as "a short, easy, and infallible method of humbling the English," and that it should therefore not be discouraged.[43]

On the other hand, William Whipple, another New Englander on the Marine Committee, was convinced that the self-serving materialism of privateering was more a curse than a blessing and that the need to nurture American virtue demanded its restraint, not its encouragement. Whipple was a strong believer in moral integrity and virtue. He warned that "nothing less than the fate of America, depends on the virtue of her sons, and if they have not virtue enough to support the most glorious cause that ever human being were engaged in, they don't deserve the blessings of Freedom."[44] Although he conceded that privateers caused considerable damage to British trade, he also maintained that even more damage would have been inflicted without them:

> ...had there been no privateers is it not probable there would have been a much larger number of Public Ships than has been fitted out, which might have distressed the Enemy nearly as much & furnished these States with necessaries on much better terms than they have been supplied by Privateers?

Whipple then declared that "no kind of Business can so effectually introduce Luxury, Extravagance and every kind of Dissipation, that tend to the destruction of the morals of people. Those who are actually engaged in it soon lose every Idea of right and wrong." By way of evidence he insisted that "no public ship will ever be manned while there is a privateer fitting out. The reason is plain: Those people who have the most influence with Seamen think it their interest to discourage the Public service, because by that they promote their own interest, viz., Privateering."[45]

---

[42]John Adams to Abigail Adams, 12 August 1776, in *NDAR*, VI, 158.

[43]Cited in Bradford, "Navies," 6.

[44]William Whipple to John Landon, 7 January 1777, in *NDAR*, VII, 856n.

[45]Whipple to Josiah Bartlett, 12 July 1778, quoted in Allen, *Naval History*, I, 48-49. Whipple's opinion was a common one in the eighteenth century. Thirty years

Outside the Marine Committee, Whipple's arguments were echoed both by national and state authorities. The navy's manning problems were shared by the various state navies as well as by the Continental Army. This prompted recommendations which generated controversy. For instance, some were prepared to restrain privateering whenever it was felt that public measures of war might benefit thereby. Manning difficulties were more widespread, and judged to be more serious, than were shortages of materiel such as shot, cannon, and gunpowder. William Vernon, a member of the Eastern Navy Board, urged the Marine Committee to declare an embargo on all shipping throughout the colonies until the navy could be completely manned: "The infamous practice of seducing our Men to leave the ships...will make it impossible ever to get our ships ready to Sail in force."[46] Commodore Hopkins had pleaded for such an embargo as early as 1776, and late in that year Massachusetts actually imposed an embargo on local shipping.[47] Such measures were guaranteed to stir up as much controversy as the problem they were intended to alleviate. There was little evidence to suggest that embargoes would have the desired effect, while the suspension of privateering would cause numerous opportunities for damaging British trade to be missed.[48] Worst of all, embargoes were "Shackles upon Trade" which violated the principle of freedom, for which the Americans were fighting, in the name of the public good.[49] Both sides of the privateering issue were therefore quick to appeal to principle.

The collapse of the Hopkins expedition did not put an end to American efforts to destroy the Newfoundland fishery. The American frigates *Hancock* and *Boston* sailed onto the Banks of Newfoundland in the spring of 1777. Their mission, claimed one of the commanding officers, was "to destroy the Fishery by sinking, burning, taking or destroying all I may find, which busi-

---

earlier, Admiral Vernon had criticized privateers, claiming that "serving the public is the least part of their thoughts or inclinations. The attaining plunder from merchant ships...being the principal motive;" cited in Daniel A. Baugh, *Naval Administration in the Age of Walpole* (Princeton, 1965), 20.

[46]William Vernon to John Adams, 17 December 1776, quoted in Allen, *Naval History*, I, 49.

[47]Esek Hopkins to the Marine Committee, 19 June and 2 November 1776, in *NDAR*, V, 622-623 and VII, 17; and Acts and Resolves of the Massachusetts General Court, 26 March 1777, in *NDAR*, VIII, 203.

[48]James Warren insisted that "The amazing damage we should have done them, as well as the advantages derived to ourselves, make me execrate the policy of stopping our privateers. I always opposed it." Warren to John Adams, 23 April 1777, in *NDAR*, VIII, 405-406.

[49]John Adams to James Warren, 6 April 1777, in *NDAR*, VIII, 282.

ness I am ordered by the Congress to do."[50] In fact, they had been instructed to place themselves and the frigate *Raleigh* at the service of the states of New Hampshire and Massachusetts which were prepared to help equip, outfit and man the warships in return for their assistance in clearing local waters of enemy cruisers. They were then to join Hopkins' squadron, which had been ordered on a similar mission on the coast of North Carolina. Only if Hopkins' ships had already left were the three frigates to remain in northern waters, and then only to cruise between Newport, Rhode Island, and the Banks of Newfoundland. Nothing was said about destroying the British fishery.[51]

Although these orders had been issued in October 1776, none of the frigates would be ready before May 1777. By then, much of what had been contained in the original orders was no longer relevant. Nevertheless, the Massachusetts government was still determined to have the Continental frigates clear local waters of enemy cruisers. To solve the frigates' manning problems and thereby hasten their departure, the state government had therefore imposed an embargo in December on all shipping. As it became apparent that two, but not three, of the frigates would soon be ready for sea, the authorities decided to lift the embargo for those privateers willing to serve for twenty-five days with the frigates. The privateers must have found the offer attractive. Because of the embargo, they had idled uselessly (and unprofitably) in Boston harbour. The Massachusetts government even offered to pay for their expenses. In return, each privateer agreed to pay a bond to ensure that they "comply with these terms and not to leave the Fleet."[52] Despite these precautions, neither the

---

[50]TNA/PRO, Admiralty (ADM) 1/471, Secretary In-Letters, 118-120, statement of T. Hardy, master of the banking brig *Patty*, 11 June 1777.

[51]Marine Committee to Captains John Manley, Hector McNeill and Thomas Thompson, 23 October 1776, in *NDAR*, VI, 1385; and Marine Committee to Commodore Esek Hopkins, 23 October 1776, in *NDAR*, VI, 1384. See also Marine Committee to Thompson, 21 September 1776, and Marine Committee to Cushing, 21 September 1776, in *NDAR*, VI, 933-935; Acts and Resolves of the Massachusetts General Court, 17 April 1777, in *NDAR*, VIII, 359-360; Minutes of the Massachusetts Board of War, 22 April 1777, in *NDAR*, VIII, 400; Acts and Resolves of the Massachusetts General Court, 24 April 1777, in *NDAR*, VIII, 416; and Journal of Capt. Hector McNeill, 21 May 1777, in *NDAR*, VIII, 1006.

[52]Owners of Massachusetts privateers to the Massachusetts General Court, 19 April 1777, in *NDAR*, VIII, 375-376; Report to the Massachusetts General Court of the Committee appointed to confer with Capt. Manley, 21 April 1777, in *NDAR*, VIII, 389-390; and Acts and Resolves of the Massachusetts General Court, 26 April 1777, in *NDAR*, VIII, 434-436. *American Tartar*, an unusually large and powerful privateer, paid a £2000 bond. What the others paid is not known; endorsement of 22 March 1777 by a committee of the Massachusetts General Court on the petition of the agents for the privateer ship *American Tartar*, in *NDAR*, VIII, 179-180.

privateers nor the frigates were prepared to give more than token support for the type of operation desired by the Massachusetts government. Within a few days of their departure from Boston on 21 May 1777, the nine privateers which joined the flotilla became separated from the frigates.[53] Shortly thereafter, the frigates also abandoned the mission and set their course for Newfoundland.

The disappearance of the privateers was not unanticipated. The Massachusetts government had imposed bonds on the privateers to enforce their cooperation. The bonds, however, included an escape clause permitting the privateers to leave the frigates "through absolute Necessity," a right which they were quick to exercise. Nor would the privateers have been confident in the leadership of the expedition. A breach had existed for some time between the frigate captains, John Manley and Hector McNeill, with the result that the success of the mission had been in doubt even before the ships set sail. In March 1777, James Warren had likened the two officers to "the Jews and Samaritans [who] will have no connections or intercourse: they will not sail together."[54] The privateers, whose priority was a profitable voyage, would not have been enthusiastic about entrusting themselves to a discontented leadership. Besides, the profit-orientation of the privateers was rarely compatible with the priorities of public service, as William Whipple had warned. In this particular instance, at least, the Massachusetts government evidently concurred; that was why the bonds had been imposed. Finally, the privateers must have known or sensed that the frigates themselves were not committed to their mission but intended instead to mount an attack on Newfoundland. The privateers had never agreed to be part of such an operation. Consequently they wasted little time in separating from the company of the warships.

As for the frigates, it is quite likely that they had cooperated with the Massachusetts General Court in order to expedite the always difficult task of fitting out. Upon their agreeing to the request of the state government, such essentials as gunpowder had quickly been made available.[55] But after patrolling the coastal waters of New England for only ten days, the frigates headed north for the waters off Newfoundland. It is difficult to conceive how they could

---

[53]Gardner, *Naval History*, I, 203. By the time *American Tartar* was captured later that summer, it had taken eight prizes, all in the North Sea or off the Shetland Islands. None were taken in North American waters, its ostensible hunting ground when she set out on her cruise; TNA/PRO, ADM 1/2, 120, No. 13, Capt. John McBride to Philip Stephens, 7 November 1777.

[54]James Warren to John Adams, 23 March 1777, in *NDAR*, VIII, 184; see also Gardner, *Naval History*, I, 203.

[55]Maj. Gen. William Heath to Langdon, 8 April 1777, in *NDAR*, VIII, 293; and W. Gardner to Capt. W. Furnel, 11 April 1777, in *NDAR*, VIII, 318.

expect to cause the fishery there serious injury with only two warships. When, after capturing a dozen or so banking vessels, they succeeded in taking HMS *Fox* (28), a frigate belonging to the Newfoundland station, the Americans headed home. Apparently they were overcome by this unexpected triumph. Then, as they neared Massachusetts, they ran afoul of British warships. *Hancock* was captured, *Fox* was retaken and only *Boston* escaped. The destruction of a dozen banking vessels could not disguise the fact that this second American attempt to attack the British fishery at Newfoundland had also failed.[56]

This was to be the last deliberate attack by ships of the Continental Navy upon the fishery. With France entering the war, the Americans had to tread more carefully in planning offensive operations. The new allies were suspicious of each other's intentions regarding Newfoundland, and neither was willing to antagonize the other by pursuing too aggressive a policy against the British fishery.[57] Individual privateers continued to cruise in Newfoundland waters, as did state cruisers, but their aim was always to intercept the transatlantic trade. For instance, in 1780, the Massachusetts state navy ship *Protector* had been ordered "to the Latitude 38 & 39, which we apprehend will be in the track of Vessels bound from Europe to New York or Carolina...then Steer to...Cape Race in Newfoundland, which Cape we imagine most of the Vessels bound from England to Canada always endeavour to make." It was under such directions that privateers and state cruisers that year achieved one of their most spectacular successes of the war, capturing a substantial number of ships belonging to a convoy destined for Québec.[58] A year earlier, American cruisers had intercepted the Jamaica fleet on the Banks of Newfoundland and captured

---

[56]TNA/PRO, ADM 1/471, 118-120, statement of Hardy, 11 June 1777; Provincial Archives of Newfoundland and Labrador (PANL), GB 2/1, Royal Engineers, Correspondence, 1774-1777, 102-103, Capt. R. Pringle to Board of Ordnance, 20 June 1777; and TNA/PRO, ADM 1/1611, No. 2, Sir George Collier (Halifax, HMS *Rainbow*) to Stephens, 12 July 1777. The cruise of *Hancock* and *Boston* is described in some detail in Allen, *Naval History*, I, 202-216; while the description in Bradford, "Navies," 15 is much briefer, it recognizes that the privateers were never really a part of the frigates' plan to cruise up to the Newfoundland banks.

[57]Dallas Irvine, "The Newfoundland Fishery: A French Objective in the War of American Independence," *Canadian Historical Review*, XIII, No. 3 (1932), 281-283.

[58]James P. Baxter (ed.), *Documentary History of the State of Maine, Containing the Baxter Manuscripts* (Boston, 1914), 253, Board of War (Boston) to John Foster Williams, commander of *Protector*, May 1780; TNA/PRO, ADM 1/486, 406, Admiral Mariot Arbuthnot to Stephens, 20 August 1780; and *Warren-Adams Letters*, II, 141, James Warren to John Adams, 12 October 1780. Ironically, because no Newfoundland-bound ships were taken, Keith Matthews declared 1780 to be the year in which "the American threat was finally mastered" in Newfoundland; Matthews, "History," 480.

eleven ships.[59] In neither instance was the British fishery there a target. While American privateers and public warships did occasionally attack fishermen and fishing outports, the fishery was remarkably untouched by them when the war finally ended in 1783.[60]

What conclusions can then be drawn about the failure of the American revolutionaries to strike at the British fishery at Newfoundland? In large measure, the answer seems to rest with the way in which the Americans carried their struggle to sea. Much of the American effort was poured into privateering, which was less significant than was once believed. As one historian explained, the privateers were "a festering and annoying thorn in the British Lion's paw, but they were in no manner the decisive factor in the outcome of the war." British reaction to the privateers was a "cry of rage rather than of pain."[61] The Newfoundland fishery provides an excellent demonstration of the limitations of *guerre de course.* Although its economic and strategic importance was undeniable, and despite the fact that it was lightly defended, the fishery did not make a worthwhile target for the privateers. Those of them which cruised Newfoundland waters did so in search of the Canada or Caribbean trade which always crossed the Banks on its transatlantic passage. Fishing vessels were attacked, but more as targets of opportunity. Privateers did attack the Newfoundland trade which, in contrast to the Newfoundland fishery, provided profitable and therefore attractive targets. The privateers, however, preferred to attack this trade in European waters, where shipping routes converged and made targets easier to find, and where captured vessels and cargoes could be quickly disposed in the market ports of friendly European countries like Spain and France. The greatest injury inflicted upon the Newfoundland fishery by the war was caused by the Royal Navy, when its rapid expansion, beginning in 1778, interfered with the fishery's supply of labour.[62]

Since privateers were unlikely to threaten the fishery with destruction, responsibility for such a strategy was left to public warships. State navies were usually employed in the defence of local waters. While state warships were occasionally sent into Newfoundland waters, they generally were like the privateers, hunting transatlantic convoys of merchantmen and supply ships whose cargoes would support or finance the revolutionary war effort. Consequently,

---

[59]Fowler, *Rebels*, 101-102.

[60]The problem of the fishery's defence during this war is examined more fully in Janzen, "Royal Navy," 28-48.

[61]Prof. Sidney Morse, cited in Morgan, "American Privateering," 11.

[62]Olaf Uwe Janzen, "Newfoundland and British Maritime Strategy during the American Revolution" (Unpublished PhD thesis, Queen's University, 1983), 248-249; and Matthews, "History," 475, 480-481.

the only noteworthy efforts to destroy the fishery came from the national navy. The decision in 1775 to establish a Continental Navy was supposed to have given Patriot leaders not only the means by which to defend American commerce at sea but also the weapon with which to carry the war to the enemy by attacking targets of strategic and economic value to England, such as the fishery at Newfoundland. But the Continental Navy was an imperfect weapon, as the abortive raid on the fishery demonstrated. The construction, outfitting and manning of Continental warships was extremely slow and inefficient. This was perhaps the unavoidable consequence of an infant and inexperienced administration, as well as of the necessity to compete with too many private and local public warships for too few men and materiel. Thus, the Hopkins expedition never did manage to sail against the Newfoundland fishery, while the frigates *Hancock* and *Boston* required nearly seven months of effort after receiving their instructions before they were able to put out to sea. Still another problem was the lack of a strong sense of discipline among the navy's officer corps; one historian described them as "fiercely independent prima donnas."[63] It is a description which suits Commodore Hopkins, who ignored his instructions to clear Chesapeake Bay of enemy cruisers in order to attack the Bahamas. It also suits captains Manley and McNeill, who were exceeding their instructions when they attacked the Newfoundland bank fishery.

Finally, and perhaps most seriously, the Continental Congress seemed uncertain about what the navy was supposed to do. Any navy must fulfil a variety of missions, but for those missions to serve best the national interest, they must conform to a national naval policy which in turn must reflect a consensus among the national leaders. This the Continental Navy lacked. Too often, orders were given but allowed to fall by the wayside or not followed through. Thus, the plan to attack the British fishery at Newfoundland in 1776 was well-conceived. Yet it was poorly executed and never revived. And even though the effectiveness of the navy suffered by its inability to compete with privateers for men and materiel, the Marine Committee never was able to develop a solution to the problem which was acceptable to all its members. Perhaps the most perceptive explanation for the navy's limitations was that of Cotton Tufts, a Boston physician and Patriot:

> It has been our Misfortune to be plunged into more Business than we could possibly conduct with any Degree of Clearness and to enter upon new Business before we had finished old. Necessity has often compelled us to this. Wisdom points out what is profitable and necessary to be done, but Means are not always at hand. In this Case We must pass on to what is

---

[63]Fowler, *Rebels*, 234.

practicable and content ourselves with a Lesser Good where we cant obtain a greater.[64]

The inability of the Americans to mount an effective offensive against the British fishery at Newfoundland at any time during the war suggests that Tufts was right.

---

[64]Cotton Tufts to John Adams, 24 April 1777, cited in Allen, *Naval History*, I, 416.

Appendix

This is a series of four oil paintings by the British marine artist Francis Holman for the captain of HMS *Flora*, signed lower left "FxHolman 1799." Executed from eyewitness accounts, the series depicts the naval engagement of 7-8 July 1777 between the Continental frigate *Hancock*, the prize ship *Fox*, the Continental frigate *Boston* and the British warships HMS *Flora* and HMS *Rainbow* off Cape Sable. (Courtesy of the Peabody Essex Museum, Salem, Massachusetts.)

This is the first of a series of four oil paintings completed in 1799 by the British marine artist Francis Holman for John McBride, who commanded HMS *Flora* when she and HMS *Rainbow*, Capt. Sir George Collier, encountered the Continental frigates *Hancock* and *Boston* with their prizes, the former British frigate *Fox* and an unidentified sloop. Left to right: *Hancock*, *Fox*, *Boston*, *Flora*, the sloop (which had been set ablaze) and *Rainbow*.

Flora exchanged fire first with *Boston*, Capt. Hector McNeill, who broke away from the engagement to make repairs to his ship. *Flora* then bore down upon *Fox* and *Hancock*, Capt. John Manley, whose resistance to the British frigate was cut short by the approach of *Rainbow*. The bitter enmity between the two American frigate captains would later lead to accusations by Manley that McNeill had stood off while *Hancock* faced the British alone. McNeill insisted that the repairs needed by *Boston* gave him no choice. Left to right: *Fox*, *Hancock*, *Flora*, *Boston* and *Rainbow*.

With *Boston* out of the fight and *Fox* too lightly manned by her prize crew to provide assistance, *Hancock* was in no position to engage both *Flora* and *Rainbow*. The Americans therefore scattered in different directions. *Flora* remained intent on recapturing *Fox* while *Rainbow* went in pursuit of *Hancock*. Left to right: *Hancock*, *Rainbow*, *Flora*, *Boston* and *Fox*.

Fox was quickly re-taken by *Flora*. After a thirty-nine-hour chase, *Hancock* was taken by *Rainbow*. Only *Boston*, with no pursuer, managed to escape. *Hancock* and *Fox* were subsequently carried into Halifax, Nova Scotia. Left to right: *Hancock*, *Rainbow*, *Flora*, *Fox* and *Boston*.

Printed and bound by CPI Group (UK) Ltd, Croydon, CR0 4YY

27/10/2024

14580407-0005